Eastern Snow Conference 2019

Fairlee, Vermont, USA
4-6 June 2019

Editors:

Krystopher Chutko
Eli Deeb

ISBN: 978-1-7138-0354-6

Printed from e-media with permission by:

Curran Associates, Inc.
57 Morehouse Lane
Red Hook, NY 12571

Some format issues inherent in the e-media version may also appear in this print version.

Eastern Snow Conference
C/O Dr. Krystopher Chutko
117 Science Pl-Dept. Geography
Saskatoon, Sk, Canada S7N 5C8

https://www.easternsnow.org/

Additional copies of this publication are available from:

Curran Associates, Inc.
57 Morehouse Lane
Red Hook, NY 12571 USA
Phone: 845-758-0400
Fax: 845-758-2633
Email: curran@proceedings.com
Web: www.proceedings.com

Proceedings of the
76th ANNUAL

EASTERN SNOW CONFERENCE

4-6 June 2019

Lake Morey Resort
Fairlee, Vermont, USA

CONTENTS

Session #3: From Microstructure to Bulk Properties

Poster Session

Town Hall / Panel: Future Snow Satellite Missions

FOREWORD

This proceedings volume contains papers presented at the 76th Eastern Snow Conference (ESC) held 4-6 June 2019 in Fairlee, Vermont, USA. The meeting featured oral and poster sessions on recent advances in remote sensing, remote sensing theory, applications of remote sensing ranging from the ice sheets to the tropics, snow and ice processes, and hydroclimatology. The meeting also included a panel discussion on the future of snow satellite missions.

The ESC is a joint United States and Canadian forum for discussing recent work on operational, applied and scientific issues related to snow and ice. It also retains an increasing interest as a symposium where novel approaches to cryospheric science of international significance are presented. The ESC has published an annual series of proceedings since 1952. Typical topics include studies of snow and ice as materials, snow removal, meteorological forecasting, river ice control, snow hydrology, snow chemistry, glaciology, remote sensing of snow and ice, and snow ecology. Membership in the ESC is open to all interested individuals and corporations. Additional copies of the current proceedings and all back issues can be obtained from the Secretary. Additional information about the Eastern Snow Conference may be found at: http://www.easternsnow.org/.

We thank members of the ESC executive committee for their continued work toward making each annual meeting a success. This includes, but is not limited to, the work of the Research Committee who evaluate all student presentations and recommend awardees for the three student prizes.

Logistical support for the 2019 Eastern Snow Conference was provided by the Cold Regions Research and Engineering Laboratory (CRREL) in Hanover, NH. Special thanks to Renee Melendy, CRREL program specialist, for assisting with the meeting.

The 2019 meeting of the Eastern Snow Conference and these Proceedings were made possible in part by Corporate membership from Campbell Scientific Canada, Geonor, Gradient Wind Engineering, and Hoskin Scientific Ltd.

We look forward to seeing you at the 77th Eastern Snow Conference meeting to be held 9–11 June 2020 at York University in Toronto, Ontario, Canada.

Krystopher Chutko and Eli Deeb
ESC Proceedings Co-Editors
University of Saskatchewan and CRREL

STATEMENT OF PURPOSE

The Eastern Snow Conference (ESC) is a joint Canadian/US organization founded in the 1940s, originally with members primarily from eastern North America. Our current members are scientists, snow surveyors, engineers, technicians, professors, students, and operational and maintenance professionals from North America, the United Kingdom, Japan, and Germany. There is a western counterpart to the ESC, the Western Snow Conference (WSC), which also is a joint Canadian/US organization. Every fifth year, the ESC and the WSC hold joint meetings.

The Eastern Snow Conference is a forum that brings the research and operations communities together to discuss recent work on scientific, applied, and operational issues related to snow and ice. The location of the conference alternates yearly between the United States and Canada, and attendees present their work by either giving a talk or presenting a poster. Most resulting papers are reviewed, edited, and published in our yearly *Proceedings of the Eastern Snow Conference*. In recent years, the ESC meetings have included sessions on snow physics, winter survival of animals, snow and ice loads on structures, river ice, remote sensing of snow and ice, and glacier processes. Volumes of the *Proceedings* can be found in libraries throughout North America and Europe, and the papers are also available through the National Technical Information Service (NTIS) in the United States and CISTI in Canada.

* * * * * * * * *

Le Colloque sur la neige région Est (ESC) est une organisation américaine-canadienne fondée dans les années ’40 et dont les membres provenaient à l’origine principalement de l’est de l’Amérique-du-Nord. Actuellement, les membres, qu’ils soient chercheurs, techniciens, ingénieurs, professeurs, étudiants, et spécialistes des services d’exploitation et d’entretien, viennent non seulement d’Amérique-du-Nord, mais aussi du Royaume-Uni, du Japon, et d’Allemagne. Le Colloque sur la neige-région Ouest (WSC), aussi une organisation américaine-canadienne, est l’homologue de l’ESC pour l’ouest nord-américain. Tous les cinq ans, l’ESC et la WSC organisent des réunions en commun.

L’ESC est un forum qui rassemble chercheurs et responsables des services d’exploitation pour discuter des travaux récents sur les problèmes scientifiques, opérationnels, ou autres dus à la neige et à la glace. Le site de cette réunion annuelle alterne entre les États Unis et le Canada. Les participants y présentent les résultats de leurs travaux par des communications orales ou au moyen d’affiches. Ces communications, une fois revues et éditées, sont publiées dans les *Annales de l’ESC*. Dans les années récentes, les réunions de l’ESC ont inclus des sessions sur la physique de la neige, la survie hivernale de la faune, les forces exercées par la neige et la glace sur les structures et les bâtiments, la glace de rivière, la télédétection de la neige et de la glace, et les processus glaciaires. Les *Annales de l’ESC* sont accessibles dans la plupart des bibliothèques scientifiques d’Amérique-du-Nord et d’Europe. Des copies d’articles peuvent être obtenues du National Technical Information Service (NTIS) aux États Unis et son équivalent au Canada, le CISTI.

EXECUTIVES FOR THE 2018-19 76th EASTERN SNOW CONFERENCE

PRESIDENT
George Riggs
Gambrills
Maryland, USA

PAST PRESIDENT
Alexandre Roy
Trois-Rivières
Québec, Canada

VICE-PRESIDENT AND PROGRAM CHAIR
Eli Deeb
Hanover
New Hampshire, USA

SECRETARY-TREASURER (Canada)
Krystopher Chutko
Saskatoon
Saskatchewan, Canada

Alexandre Langlois
Sherbrooke
Québec, Canada

SECRETARY-TREASURER (USA)
Ken Rancourt
Conway
New Hampshire, USA

ESC PROCEEDINGS EDITORS
Krystopher Chutko
Saskatoon
Saskatchewan, Canada

George Riggs
Gambrills
Maryland, USA

***PHYSICAL GEOGRAPHY* EDITORS**
Mauri Pelto
Dudley
Massachusetts, USA

STEERING COMMITTEE
Craig Smith (Chair)
Saskatoon
Saskatchewan, Canada

Sean Helfrich
Suitland
Maryland, USA

Steve Howell
Toronto
Ontario, Canada

Joan Ramage
Bethlehem
Pennsylvania, USA

David Robinson
New Brunswick
New Jersey, USA

Laura Thomson
Kingston
Ontario, Canada

Carrie Vuyovich
Hanover
New Hampshire, USA

RESEARCH COMMITTEE

Barton Forman (Chair)
College Park
Maryland, USA

Laura Brown
Toronto
Ontario, Canada

James Brylawski
Augusta
New Jersey, USA

Josh King
Toronto
Ontario, Canada

Benoit Montpetit
Ottawa
Ontario, Canada

Tara Troy
Victoria
British Columbia, Canada

Sam Tuttle
South Hadley
Massachusetts, USA

WEBMASTER

Vincent Sasseville
Sherbrooke
Quebec, Canada

LOCAL ARRANGEMENTS

Eli Deeb
Hanover
New Hampshire, USA

Joan Ramage
Bethlehem
Pennsylvania, USA

Carrie Vuyovich
Gambrills
Maryland, USA

THE PRESIDENT'S PAGE

The 76th annual meeting of the Eastern Snow Conference was held at the Lake Morey Resort in Fairlee, VT, on 4-6 June 2019. A serene country setting compared to the urban bustle of College Park, MD where last year's ESC was held. It was enjoyable to watch loons diving in the lake and to hear them calling at night. Lake Morey Resort was chosen because it is just up the road from Cold Regions Research and Engineering Laboratory (CRREL) in Hanover, NH.

The scientific program included 25 oral presentations, 36 poster presentations and a town hall/panel session on future snow satellite missions moderated by Edward Kim, and Michael Durand, with panelists, Carrie Vuyovich and Joshua King. The town hall/panel session, which was a first for an ESC that I'm aware of, had excellent participation and was a great exchange of ideas and plans between US and Canadian scientists. A wide range of research and applications concerning snow and ice were presented by ESC participants. A large number of students attended and presented their research. The ESC had a social media presence, Twitter, with many posts made by participants. The Research Committee at the Banquet recognized the following students for exemplary achievement:

- Zhibang Lv of the University of Saskatchewan, was awarded the Wiesnet Medal for best student presentation for "Assimilation of snow interception information into a cold regions hydrological model".
- Gaohong Yin of the University of Maryland, College Park, was awarded the David Hewitt Miller Best Student Poster Award for "Characterizing Snow Water Equivalent from Ground-based Observations of GPS Vertical Displacement and Model-based Hydrologic Loading Estimates".
- Alex Mavrovic of the Université de Sherbrooke, was presented the Campbell Scientific Canada prize for "Development of an open-ended coaxial probe (OECP) for snow liquid water content measurement".

We were greatly pleased to have Dr. Don Perovich, Professor of Engineering, Dartmouth College and former research scientist at CRREL deliver the Banquet Address.

Barry Goodison gave a preamble on the history and significance of the coveted Sno-foo Award. Alexandre Langlois presented the coveted Sno-foo Award to Alexandre Roy for letting slide the solemn duty of attaching award winner names to the Award.

Special thanks to Eli Deeb the Vice-President and Program Chair and his local arrangements team of Joan Ramage, Carrie Vuyovich and CRREL volunteers. Thanks to ESC Executive Committee Members Krys Chutko, Derrill Cowing and Ken Rancourt, who continue to provide outstanding leadership and were instrumental to the success of the 76th meeting. The efforts of webmaster Vincent Sasseville, and members of the steering and research committees, notably Barton Forman for revising the Wiesnet Medal guidelines to increase student interest, are also greatly appreciated. Likewise, the meeting would not have been a success without the participation and continued support of our corporate members: Campbell Scientific Canada, Geonor, Hoskin Scientific and new corporate member Gradient Wind Engineering.

Please plan to attend the 77th ESC in Toronto, Canada, early June 2020. Krys Chutko, University of Saskatchewan, is the Vice-President and Program Chair. Check the ESC website for updates and students should review the guidelines for the Wiesnet Medal, the David Hewitt Miller Best Student Poster Award and Campbell Scientific Canada prize.

Thank you for your participation and support of the Eastern Snow Conference. Best wishes for success in your field work, model development, research or whatever your contribution is to better understanding the cryosphere. When the chill of winter comes, remember to pull on your ESC hat.

George Riggs
76th President, Eastern Snow Conference

LIFE MEMBERS OF THE EASTERN SNOW CONFERENCE

The Eastern Snow Conference gratefully recognizes individuals who have made lifelong contributions to the study of snow and for their support of this organization.

JAMES FOSTER

BARRY GOODISON

GERRY JONES

JOHN METCALFE

HILDA SNELLING

DONALD WIESNET

Wiesnet Medal

The Eastern Snow Conference encourages student research through its Wiesnet Medal. This medal is presented annually to the best student paper presented at the conference.

Year	Winner	Affiliation
2019	Zhibang Lv	University of Saskatchewan
2018	Jiyue Zhu	University of Michigan
2017	Caroline Dolant	Université de Sherbrooke
2016	Syed Mousavi	University of Michigan
2015	Nicolas Leroux	University of Saskatchewan
2014	Justin Hartnett	Syracuse University
2013	Andreas Dietz	Earth Observation Center/DFD, Germany
2012	Elizabeth Burakowski	University of New Hampshire
2011	Kathryn Semmens	Lehigh University
2010	Simon von de Wall	University of Victoria
2009	Si Chen	Dartmouth College
2008	Chris Fuhrmann	University of North Carolina at Chapel Hill
2007	*Not awarded*	
2006	Y.C. Chung	University of Michigan
2005	M. Javan-Mashmool	Université du Québec à Chicoutimi
2004	J. Farzaneh-Dehkordi	Université du Québec à Chicoutimi
2003	Alexandre Langlois	Université de Sherbrooke
2002	Patrick Ménard	Université de Laval
2001	C. Tavakoli	Université du Québec à Chicoutimi
2000	*Not awarded*	
1999	S. Brettschnieder	Université du Québec à Chicoutimi
1998	Andrew Grundstein	University of Delaware
1997	Newell Hedstrom	University of Saskatchewan
1996	Suzanne Hartley	University of Denver
1995	Paul Wolfe	Wilfred Laurier University
1994	G.E. Mann	University of Michigan
1993	G. Devarennes	Université du Québec à Québec
1992	D.W. Cline	University of Colorado
1991	D. Samelson	Cornell University
1990	A.K. Abdel-Zaher	University of New Brunswick
1989	A. Giguere	McGill University
1988	Mauri Pelto	University of Maine
1987	Cameron Wake	Wilfred Laurier University
1986	Craig Allan	Trent University
1985	Robert Speck	Rensselaer Polytechnic Institute
1984	N.K. Kalliomaki	Laurentian University
1983	David Beresford	Trent University
1982	*Not awarded*	
1981	Jeffrey Patch	University of New Brunswick
1980	Bryan Wolfe	Trent University
1979	Margaret Leech	McGill University
1978	Michael English	Trent University
1977	Don McLaughlin & George Duggan	Rensselaer Polytechnic Institute
1976	Dwayne McMurter	Trent University
1975	Nigel Allan	Syracuse University
1974	*Not awarded*	
1973	Stan Mathewson	Trent University

David Miller Award

The David Miller Award is presented to the best student poster at the annual Conference.

Year	Winner	Affiliation
2019	Gaohong Yin	University of Maryland
2018	Justin Pflug	University of Washington
2017	Elizabeth Ryan	University of Maryland
2016	Oliver Wigmore	The Ohio State University
2015	Charles Papasodoro	Université de Sherbrooke
2014	Laura Thomson	University of Ottawa
2013	Nastaran Saberi	University of Waterloo
2012	Johnathan Sugg	Appalachian State University
2011	*Not awarded*	
2010	Nicolas Svacina	University of Waterloo
2009	Iliyana Dobreva	Texas A&M University
2008	Christopher Karmosky	Pennsylvania State University

Campbell Scientific Canada Prize

The Campbell Scientific Canada prize is awarded to student research showing the most innovative use of technology in the gathering of data.

Year	Winner	Affiliation
2019	Alex Mavrovic	Université de Sherbrooke
2018	*Not awarded*	
2017	Abigail Dalton	University of Ottawa
2016	*Not awarded*	
2015	*Not awarded*	
2014	Aaron Campbell	University of Waterloo
2013	*Not awarded*	
2012	Andrew Kasurak	University of Waterloo
2011	Mathieu Beaulieu	University of British Columbia
2010	*Not awarded*	
2009	*Not awarded*	
2008	Nicholas Kinar	University of Saskatchewan
2007	Gro Lilbaek	University of Saskatchewan
2006	Gernot Koboltschnig	University of Natural Resources and Applied Life Sciences, Vienna
2005	Warren Helgason	University of Saskatchewan

Session #1: Sea and Lake Ice

Session Chair: Joan Ramage

Spatiotemporal Polynya Formation Trends in the Canadian Arctic Archipelago using Sea Ice Charts from 1968 Onwards

GUILLAUME COUTURE[1], ALEXANDRE LANGLOIS[1], STEPHEN HOWELL[2], AND BENOIT MONTPETIT[3]

ABSTRACT

Arctic temperatures increase up to 1.9 times faster than the rest of the globe. This greatly affects sea ice conditions. In the Canadian Arctic Archipelago (CAA), polynyas (open water areas forming over winter) greatly contribute to the ice breakup during the melt season. Monitoring of these areas are of particular importance because an increase in size and occurrence could lead to earlier ice breakup and contribute to Arctic amplification. The main goal of this project is to analyze the spatiotemporal trends in polynya formation for the CAA. To do so, ice concentration anomalies, polynya occurrences, and opening periods were studied using the Canadian Ice Service charts. A significant increase in open water/thin ice areas was observed for the months of April, May, June and July from the mid- 2000s onwards. This rise in monthly occurrences was compared with air temperatures and wind anomalies. This showed that temperature increase was more likely to be the cause for the higher count of polynyas than winds. Lastly, a detection algorithm developed by Nemer *et al.* (2016) using fuzzy logic and discrete wavelet transform was used to segment and classify RADARSAT 1-2 images to assess its potential for automatic open water area retrievals. After testing the algorithm, some issues were noted with the classification such as open water being classified as ice and vice versa, indicating that the use of this algorithm with these images is not suitable for the variable sea ice and weather conditions of the CAA.

[1] Université de Sherbrooke, Sherbrooke, QC, Canada
[2] Environment and Climate Change Canada, Toronto, ON, Canada
[3] Environment and Climate Change Canada, Ottawa, ON, Canada

Implications of Ice Cover Characteristics for Underwater Oil Spills in the Straits of Mackinac, Michigan

GRANT GUNN[1], KELSEY NYLAND[1], VLAD TARABARA[2], MICHELLE RUTTY[3], DOUG BESSETTE[3], AND ROBERT RICHARDSON[3]

ABSTRACT

Installed in 1953, Enbridge's oil pipeline "Line 5" transports 540,000 barrels per day of light crude oil and natural gas liquids beneath the Straits of Mackinac. American portions of Line 5 start in Superior, Wisconsin, across Michigan's Upper Peninsula, then along the lake bed in the Straits of Mackinaw before terminating in Sarnia, Ontario. Recent events have raised concerns about Line 5's safety and potential for spill potential in ice-covered conditions.

Consolidated ice cover may restrict the spread of oil by shielding the released crude from wind transport and partly immobilizing the spill. Absent turbulent conditions, crude constituents should rise to the ice-water interface and pool at the ice's underside, with laboratory and field observations indicating that oil will coalesce, form a slick, and spread while resting on the ice-water interface. As a result, the roughness of the ice-water interface is of critical importance to the potential spread of oil.

In this study, we retrieve the roughness characteristics and generate topographic features of the ice underside in the Straits of Mackinac using Ground Penetrating Radar. Observed RMS roughness features of up 4cm are used to identify how ice properties impact potential oil residency time, particularly the transport of oil with respect to periods of increased flow, as well as the potential for oil weathering by emulsification. Laboratory experiments simulate oil spills under rough ice conditions and inform agent- based modeling (ABM) research and provide recommendations for under-ice oil spill remediation policy and best practices.

---

[1] Department of Geography, Environment and Spatial Sciences, Michigan State University, East Lansing, MI, USA
[2] Department of Civil and Environmental Engineering, Michigan State University, East Lansing, MI, USA
[3] Department of Community Sustainability, Michigan State University, East Lansing, MI, USA

Small-Scale Variability of Snow Properties on Sea Ice: From Snow Pits to the SnowMicroPen

STEFANIE ARNDT[1], NICOLAS STOLL[1], ARTTU JUTILA[1], AND STEPHAN PAUL[2]

ABSTRACT

Snow on sea ice alters the properties of the underlying ice cover as well as associated exchange processes at the interfaces between atmosphere, sea ice, and ocean. It contributes significantly to the sea-ice mass and energy budgets due to comprehensive seasonal transition processes within the snowpack. Therefore, several studies have shown the importance of comprehensive understanding of snow properties for large- scale estimates in the ice-covered oceans. However, field studies reveal not only a strong seasonality but especially spatial variations on floe-size scales. It is therefore necessary to locate and quantify seasonal snow processes, such as internal snowmelt, snow metamorphism, and snow-ice formation in the Arctic and Antarctic snowpack on small scales.

Doing so, we present here *in situ* observations of physical snow properties from point measurements (snow pits) and transect lines (SnowMicroPen, SMP) during recent expeditions in the Weddell Sea and off the northeastern coast of Ellesmere Island, Canada, from 2013 to 2019, covering summer and winter conditions.

Results from a case study of snow pit analyses in the Weddell Sea during austral winter reveal a high variability of snow parameters throughout the snowpack. It is shown that snow grain size dominates the spatial variability of the snow pack while snow density variability can be neglected. The additional use of the SMP allows to even quantify length-scale variabilities of snow properties in different ice regimes in both hemispheres.

Overall, results will improve our understanding of seasonal processes in the snowpack and will guide us towards upscaling approaches of vertical snow layers on Arctic and Antarctic sea ice.

---

[1] Alfred-Wegener-Institut Helmholtz-Zentrum für Polar-und Meeresforschung, Bremerhaven, Germany

[2] Department of Geography, Ludwig-Maximilians-Universität München, Munich, Germany

Non-Destructive Characterization of a Freshwater Lake Icepack using Wideband Autocorrelation Radiometry

MOHAMMAD MOUSAVI[1], ROGER DE ROO[2], KAMAL SARABANDI[1], AND ANTHONY W. ENGLAND[3]

ABSTRACT

The behavior of electromagnetic waves in homogenous media is dependent on the medium's macroscopic parameter, the relative dielectric constant. The relative dielectric constant is an electrical property of the material which changes the magnitude, phase, and direction of an applied electric field. The dielectric constant is a complex quantity. In a low-loss material, where there is no significant absorption or heat dissipation, the imaginary part of the dielectric would be nearly equal to zero, and the dielectric constant would be a real quantity.

There are many techniques in the literature to measure the dielectric constant of a material, such as the resonant technique. These methods require direct sampling of a material, which is destructive and impossible in some scenarios, such as snow on high altitude mountains. To address these issues, we introduced a novel technique for measuring the dielectric constant of a low-loss layer without the need for sample preparation. This technique inspired by a passive microwave remote sensing method, known as wideband autocorrelation radiometry (WiBAR), which measures the microwave propagation time difference of multipath microwave emission from low-loss layered surfaces, such as freshwater lake icepack. This time delay is dependent on the incident angle. By measuring the time delay at two distinct incident angles, the real part of the dielectric constant can be measured. An X-band instrument fabricated from the commercial-off-the-shelf (COTS) components are used to characterize the freshwater lake ice at the University of Michigan Biological Station.

[1] Electrical Engineering and Computer Science, University of Michigan, Ann Arbor, MI, USA
[2] Climate and Space Sciences and Engineering, University of Michigan, Ann Arbor, MI, USA
[3] College of Engineering and Computer Science, University of Michigan, Dearborn, MI, USA

Session #2: Remote Sensing of Snow

Session Chair: Jennifer Jacobs

Interactive Multisensor Snow and Ice Mapping System (IMS) Upgrades and Improvements

JOHN WOODS[1] AND SEAN HELFRICH[2]

ABSTRACT

The U.S. National Ice Center's Interactive Multisensor Snow and Ice Mapping System (IMS) produces daily snow and ice coverage products for the entire Northern Hemisphere. A certified IMS analyst interprets satellite data from multiple platforms, instrument observations, and automated snow/ice detection algorithms to provide a quality-controlled composite of snow and ice coverage, data since last observed (DSLO), and snow depth. Updated analyses are produced twice a day to ensure that the most up to date information is available for use within numerical weather prediction models. Over its 21-year history, there have been numerous updates to improve the accuracy and resolution of IMS products. The latest version of the IMS incorporates new tools and functionality, along with data sources enabling analysts to more quickly and accurately detect and analyze snow and ice.

New data sources the IMS has been successful in incorporating includes environmental observations from the recently launched GOES-16 & 17 satellites, along with Sentinel 1 A/B and NOAA-20. These next generation satellites provide analysts with an unprecedented ability to identify snow and ice in high definition with improved spatial and temporal resolution. The improvements in satellite data coupled with the deployment of new data products within the IMS, such as Arctic imagery composites, a synthetic aperture radar (SAR) ice extent product, and a corrected DSLO algorithm has contributed to a decrease in time it takes to perform an analysis, while also providing increased analysis accuracy. The improved product benefits all IMS users, including the numerical weather prediction community.

[1] NOAA / NESDIS / OSPO / NIC, Suitland, MD, USA
[2] NOAA / NESDIS / STAR, College Park, MD, USA

Creating a Roadmap for Remotely Sensed Snow Product Feasibility on a Global Scale

VICTORIA LY[1], JESSICA LUNDQUIST[1], AND MELISSA WRZESIEN[2]

ABSTRACT

Remote sensing provides a powerful tool for observing seasonal snow properties on a continuous time-scale and across local, regional, and global spatial scales. Snow products (e.g. SNODAS, GlobSnow, AMSR-E SWE) capitalize on remote sensing data to provide observations on snow cover extent, snow water equivalent, snow depth, albedo, and other snow properties for researchers and resource managers. However, the limitations of snow products as derivatives of optical and passive microwave sensors, while understood, have not been explicitly discussed and mapped. Snow products derived from optical sensors are primarily limited by tree canopy cover and cloud cover. Snow products derived from passive microwave sensors are primarily limited by complex terrain, snow depth, and snow wetness. This paper reviews snow products, identifies the limitations, and provides a first attempt to map the global limitations of optical and passive microwave sensors within Google Earth Engine. We present "snow usability masks" in a graphical user interface, where the user defines the time, spatial extent, and boundary values for parameters, like vegetation, cloud cover, snow wetness etc. The goal is to provide an open-source, easily-accessible system as a roadmap to identify areas with the highest potential of applying snow products for operational and research uses.

[1] Department of Civil and Environmental Engineering, University of Washington, Seattle, WA, USA
[2] Department of Geology, University of North Carolina at Chapel Hill, Chapel Hill, NC, USA

High Resolution Snow Depth Mapping with Unmanned Aerial Vehicle (UAV) using Structure-from-Motion (SfM) and Kinematic dGPS: Comparison of Two Methods for Arctic Application

JULIEN MELOCHE[1,2], DANIEL KRAMER[1,2], ALEXANDRE LANGLOIS[1,2], AND ALAIN ROYER[1,2]

ABSTRACT

In order to improve snow monitoring at the global scale using satellites, there is an urgent need to improve monitoring on a spatial scale smaller than satellite images to better understand the governing process controlling its spatial and temporal distribution. The main objective is thus to map snow depth using Unmanned Aerial Vehicle (UAV) adapted to Arctic conditions. 1) We produced Digital Surface Model (DSM) from 2D images using Structure-from-Motion Algorithm software. Two DSMs are needed, one with snow and one without snow. From that, the difference between the two can be computed in a GIS and produce a snow depth map. Secondly, high resolution mapping with UAV implies dGPS system for acquisition of ground control points. We decided to try a technique that uses kinematic dGPS from an antenna mounted on a snowmobile to measure dGPS points representing the snow surface. Then, by interpolating all the points using kriging interpolation into a DSM, a snow depth map was produced with the same snow off DSM used in 1) from UAV images in the summer.

[1] Université de Sherbrooke, Sherbrooke, QC, Canada
[2] Centre d'Études Nordiques, Laval, QC, Canada

Seasonal Ku-Band Radar Measurements across a Snow-Covered Tundra Basin

JOSHUA KING[1], CHRIS DERKSEN[1], BEN MONTPETIT[2], AND PAUL SIQUEIRA[3]

ABSTRACT

Current satellite observation systems are unable to estimate terrestrial snow water equivalent (SWE) at the spatial or temporal scales necessary to advance operational climate services or numerical weather prediction. Insensitivity to basin-scale changes in snow mass, limited spatial coverage and poor temporal revisit are amongst the reasons a novel space borne observation concept has become a priority. To address this gap, Environment and Climate Change Canada (ECCC), the Canadian Space Agency (CSA), and international partners, are developing a dual-frequency (17.2 and 13.5 GHz) moderate resolution (250 m) radar mission concept for global monitoring of terrestrial snow mass. As part of the mission's science activities, a coordinated airborne, satellite and *in situ* campaign to evaluate multi-frequency radar interactions with snow, vegetation and soils in the tundra was completed during the winter of 2019. An airborne 13.5 GHz interferometric synthetic aperture radar (InSAR) deployed within the Trail Valley Creek (TVC) research basin (Northwest Territories, Canada) on three occasions (December 2018, January 2019, March 2019) to characterize these interactions. Distributed snow property measurements including SWE and microstructure were completed during each flight campaign to evaluate spatiotemporal influence. Bi-monthly RadarSAT-2 and TerraSAR-X imagery was acquired to quantify variations in vegetation and soil background contributions. The acquired snow property measurements and satellite-derived background fields were used to parametrize the Snow Microwave Radiative Transfer (SMRT) model. Output of the forward model parametrizations allows decomposition of observed backscatter diversity within the TVC domain.

[1] Environment and Climate Change Canada, Climate Research Division, Toronto, ON, Canada
[2] Environment and Climate Change Canada, Landscape Science and Technology Division, Ottawa, ON, Canada
[3] Department of Electrical and Computer Engineering, University of Massachusetts, Amherst, MA, USA

Assimilation of Snow Interception Information into a Cold Regions Hydrological Model

ZHIBANG LV[1] AND JOHN W. POMEROY[1]

ABSTRACT

Snow interception is a crucial hydrological process in cold regions needleleaf forests, but it is rarely measured directly. Indirect estimates of snow interception can be made by measuring the difference in the increase in snow accumulation between the forest floor and a nearby clearing over the course of a storm. Pairs of automatic weather stations with acoustic snow depth sensors provide an opportunity to estimate this, if snow density can be estimated reliably. To find an approach to assimilate snow depth measurements estimated snow interception, three approaches for estimating fresh snow density were investigated: weighted post-storm increments of density from the physically based Snobal model, fresh snow density empirically estimated from air temperature (Hedstrom and Pomeroy, 1998), and fresh snow density empirically estimated from air temperature and wind speed (Jordan *et al.*, 1999). Automated snow depth measurements from adjacent forest and clearing sites and estimated snow densities were used to determine snowstorm snow interception at a subalpine forest in the Canadian Rockies. Then the estimated snow interception was assimilated into the physically based, flexible, modular Cold Regions Hydrological Modelling platform (CRHM) driven by Global Environmental Multiscale (GEM) model forcing data using the Ensemble Kalman Filter. Interception determined using density estimates from the Hedstrom-Pomeroy equation agreed best to the observations from a weighed, hanging tree lysimeter. Assimilating snow interception information from automatic snow depth measurements improved the modelled snow interception timing and magnitude by 7% and 13%, respectively; its accuracy was close to that simulated using locally observed meteorological data.

[1] Centre for Hydrology, University of Saskatchewan, Saskatoon, SK, Canada

Session #3: From Microstructures to Bulk Properties

Session Chair: Michael Durand

Characterization of Snow, Firn and Ice

IAN BAKER[1]

ABSTRACT

In this presentation we give an overview of techniques used to characterize the microstructures of snow, firn (multi-year snow) and ice found in both cold regions and in polar ice sheets. These techniques include: transmission electron microscopy, synchrotron-based X-ray topography, cold-stage scanning electron microscopy coupled with energy dispersive X-ray spectroscopy, electron channeling patterns, and electron backscatter patterns; cold stage confocal scanning optical microscopy coupled with Raman spectroscopy; and micro X-ray computed tomography. The capabilities and information obtainable along with the limitations and challenges of each technique will be discussed. Examples of each technique will be presented, and future technique development will be discussed.

This work was sponsored by the National Science Foundation Arctic Natural Science grant 1743106.

[1] Thayer School of Engineering, Dartmouth College, Hanover, NH, USA

Towards a New Theory of Snow Friction

JAMES H. LEVER[1], SUSAN TAYLOR[1], GARRETT R. HOCH[1], AND EMILY ASENATH-SMITH[1]

ABSTRACT

The mechanics of snow friction are central to competitive skiing, safe winter driving, avalanche dynamics, and efficient Polar sleds. For nearly 80 years, prevailing theory has postulated self-lubrication: dry-contact sliding warms snow-grains to the melting point, and further sliding produces melt-water that lubricates the interface. We recently published micro-scale interface observations that contradicted this explanation: contacting snow grains abraded and did not melt under a polyethylene slider, despite low friction values. We obtained coordinated infrared, visible-light, and scanning-electron micrographs that confirm that the evolving shapes observed during our tribometer tests are contacting snow grains polished by abrasion, and that the wear particles can sinter together and fill the adjacent pore spaces. Furthermore, dry-contact abrasive wear reasonably predicts the evolution of snow-slider contact area and sliding-heat-source theory confirms that contact temperatures would not reach 0°C during our tribometer tests. Importantly, published measurements of interface temperatures also indicate that melting did not occur during field tests on sleds and skis. We postulate that abraded ice crystals form a dry-lubricant layer that makes contacting snow-grains slippery and are currently undertaking additional observations and theoretical analyses to assess this hypothesis.

[1] Engineer Research and Development Center, Cold Regions Research and Engineering Laboratory, Hanover, NH, USA

The Relationship Between Temperature and Strength in High Density Polar Snow

GEORGE L. BLAISDELL[1] AND TERRY MELENDY[1]

ABSTRACT

Operating and maintenance decisions for the Phoenix deep-snow runway are based on index strength measurements. In allowing the operation of C-17, C-130, and several commercial aircraft types, Phoenix represents essential infrastructure for the National Science Foundation's U.S. Antarctic Program. To optimize the brief austral summer field season as well as use of military airframes, the Program's flight schedule is established and coordinated with many agencies and research programs well in advance of deployment. To ensure a high mission-completion-to-plan ratio, a runway strength predictive capability, by aircraft type, is required.

Having monitored many environmental and snow properties at the Phoenix runway site over the 30 months since its establishment, we show that near-surface ambient temperature is a reliable indicator of snow pavement strength, even at depth, provided that the snow pavement system is properly maintained (density > 0.6 g/cm3; albedo > 88%).

We show that cone penetrometer index strength ranges from 50 at temperatures near 0°C to over 250 when air temperature is below -40°C (values over about 70 are necessary for C-17 operations). Further, austral summer seasons air temperature records at the Phoenix site show a consistent and predictable rise, plateau and fall, allowing prediction of safe operating time periods by aircraft type.

[1] U.S. Army Corps of Engineers, Cold Regions Research and Engineering Laboratory, Hanover, NH, USA

Arctic Snow Modeling with a New Parameterization of Crocus to Improve Vertical Density Stratification and Soil Temperature Simulations

CÉLINE VARGEL[1,2,3], ALAIN ROYER[1,2], GHISLAIN PICARD[3], ISABELLE GOUTTEVIN[4], AND MARIE DUMONT[4]

ABSTRACT

Arctic snowpacks often exhibit a strong vertical density stratification. Due to the high wind speed conditions that are often met in these regions, hard wind slabs form on the top of the snow cover, whereas depth hoar, with a low density, develops at the bottom. On the contrary, alpine snowpack often evolved toward vertical density stratification inverse to that of the Arctic snowpack. Detailed snowpack models were first developed to study alpine snowpack for avalanche forecasting and thus fail in simulating the density profile of arctic snowpacks. In addition, snow is an important insulator for soils of the Arctic regions. As the thermal conductivity of snow is directly linked to its density, the soil temperature is also affected by large uncertainties in simulations. In Arctic areas, soil temperature is a key parameter, which is linked to e.g. permafrost evolution and ecosystems. Here, we implement a new parameterization of some physical processes in the detailed snowpack model Crocus to better reproduce the vertical density stratification of snow and thus the simulated soil temperature. The wind compaction of snow was increased and the model takes vegetation height into account which stops the compaction when the snow is protected. Also, the snow thermal conductivity from Sturm *et al.* (1997) was implemented and used which differs from the standards options in Crocus. Results show significant improvements in density and soil temperature at three evaluation sites over several years: Cambridge Bay (Nunavut, Canada), Bylot Island (Nunavut, Canada) and Samoylov (Siberia). The RMSE between density observations and simulations is reduced by 29% with this new Crocus parameterization, compared to the default parameterization. Similarly, the simulated soil temperature with the default parameterization yields a mean bias of 9.2 K, whereas the new parameterization yields a mean bias of 2.4 K. Correlations between measurements and observations are also increased with the new parameterization. This will improve the snow and soil evolution analysis in high northern latitude over time, using this new Crocus version driven by a re-analysis dataset.

[1] Centre de Recherches et d'Applications en Télédétection (CARTEL), Université de Sherbrooke, QC, Canada

[2] Centre d'Études Nordiques, Québec, QC, Canada

[3] Institut des Géosciences de l'Environnement, Université Grenoble Alpes-CNRS, Grenoble, France

[4] Iniversité Grenoble Alpes, Université de Toulouse, Météo-France, CNRS, CNRM, Centre d'Etudes de la Neige, Grenoble, France.

Wide Variety of Techniques for Field Measurements of Snow Strength

S. SHOOP[1], W. WIEDER[1], AND B. ELDER[1]

ABSTRACT

During the winter of 2018 field experiments were conducted to assess the mechanical properties of virgin, groomed and compacted snow. These strength measurement techniques assessed the bearing and shear capacity of the snow, or a combination thereof. Many of the methods were adapted from those used in soil and pavement assessments and could be related to California Bearing Ratio, and others were techniques specifically designed for snow characterization (Rammsonde, Russian snow penetrometer, CTI penetrometer, Yamaha drop cone). The results illustrate typical values and ranges for the strength of different types of snow surfaces, and the applicability or effectiveness of the different tests to specific snow conditions.

[1] U.S. Army Corps of Engineers Cold Regions Research and Engineering Laboratory, Hanover, NH, USA

Poster Session

Session Chair: Barton Forman

Spatial and Temporal Patterns of Snowmelt in the Red River of the North Basin using Enhanced Resolution Passive Microwave Data

MARISSA J. TORRES[1], CARRIE VUYOVICH[2], AND MARINA REILLY-COLLETTE[1]

ABSTRACT

The Red River of the North Basin (RRB), bordering eastern North Dakota and western Minnesota, is vulnerable to frequent snowmelt floods due to its flat terrain and low permeability soil. Advancements in passive microwave remote sensing have produced the Calibrated Enhanced-Resolution Passive Microwave Brightness Temperature (CETB), a NASA MEaSUREs product, which provides a high-resolution record of snow mass and snowmelt properties. This dataset, in addition to ground observations and modeled SWE estimates, may provide valuable information to hydrologic forecasters.

We present a regional spatial and temporal analysis of CETB SWE melt patterns in the RRB for snow years 2004-2011. We estimated CETB SWE using a simple empirically based algorithm and compared the spatial patterns to SNODAS modeled SWE estimates. Both CETB and DAV data were evaluated temporally with hydro-meteorological data (e.g., discharge, soil and air temperature) to assess the timing of melt onset in several sub-basins within the region. We identified dates of melt using common DAV melt criterion and dates of the spring pulse in streamflow using the cumulative departure from the mean. We further compared the passive microwave SWE estimates with MODIS snow covered area to confirm regional melt patterns. Preliminary results indicate a lag time as early as two days between initial melt detection and an increase in discharge. Additional observations will be presented.

[1] U.S. Army Corps of Engineers Cold Regions Research and Engineering Laboratory, Hanover, NH, USA
[2] NASA Goddard Space Flight Center, Greenbelt, MD, USA

How Enhanced-Resolution Brightness Temperatures are Improving Algorithms for Snow Water Equivalent and Melt Onset

M.J. BRODZIK[1], D.G. LONG[2], M.A. HARDMAN[1], J.M. RAMAGE[3], R.L. ARMSTRONG[1], AND R. KELLY[4]

ABSTRACT

Funded by NASA MEaSUREs, we have reprocessed the entire record of gridded SMMR, SSM/I-SSMIS and AMSR-E brightness temperatures using the radiometer version of Scatterometer Image Reconstruction (rSIR). Image reconstruction algorithms can be tuned to reduce noise or improve spatial resolution but cannot do both. Our Calibrated Enhanced-Resolution Brightness Temperature (CETB) Earth System Data Record (ESDR) includes conventional, low-noise, images at 25 km resolution, as well as enhanced- resolution images at up to 3.125 km. Input swath data comprise the newly available CSU Fundamental Climate Data Record (FCDR), with the entire, cross-calibrated SSM/I-SSMIS record from 10 sensors, some of which have never before been produced in gridded form. While these passive microwave sensors provide a 40-year observation record, previous algorithms to derive snow water equivalent and melt onset have been confounded by mixed-pixel effects in some regions. This restricted useful applications to locations distant from land/water boundaries, and regions with low topographic relief. The CETB data now provide new opportunities to revisit analysis in mountainous regions and near coastlines. We briefly describe the CETB data, and include promising examples demonstrating snow water equivalent and melt onset algorithm improvements, due to the improved spatial resolution of the CETB data.

[1] University of Colorado, Boulder, CO, USA
[2] Brigham Young University, Provo, UT, USA
[3] Lehigh University, Bethlehem, PA, USA
[4] University of Waterloo, Waterloo, ON, Canada

Testing Calibrated Enhanced Resolution Brightness Temperature (CETB) to Detect Significant Events in Lake Ice Formation and Evolution on Large Northern Lakes

JOAN M. RAMAGE[1], MARY J. BRODZIK[2], MOLLY A. HARDMAN[2], AND DAVID G. LONG[3]

ABSTRACT

Calibrated, Enhanced-Resolution Brightness Temperatures (CETB) Earth System Data Records (ESDR) records of Advanced Microwave Spectroradiometer for EOS (AMSR-E) data provide an opportunity to evaluate freeze–thaw dynamics on northern lake systems. CETB data were created using the radiometer version of the Scatterometer Image Reconstruction (rSIR) technique. These enhanced-resolution data are 64 times finer spatial resolution at 36 GHz frequencies (3.125 km pixels) than the historical 25-km data products. They provide significant improvement in the ability to distinguish finer spatial patterns, including lake margins and spatiotemporal variations across large lakes. CETB products include previously unavailable statistical variables, which capture information about the state of dynamic surfaces, such as formation and development of lake ice, snow melt, and transition to open water. Separation between the minimum and maximum rSIR daily brightness temperature is significant in heterogeneous landscapes undergoing changes. We calculate the spatial standard deviation (SD) of rSIR data and compare the spatial and temporal variation to other factors such as temperature, ice characteristics, and other sensors to determine if this is a robust indicator of environmental transitions. We investigate patterns of high-resolution spatial variation and test the ability to use the spatial standard deviation to identify significant events in lake ice formation and evolution. These ideas are explored for sample large northern lakes, such as Great Slave Lake in the Northwest Territories, Canada.

[1] Earth and Environmental Sciences, Lehigh University, Bethlehem, PA, USA
[2] National Snow and Ice Data Center, University of Colorado, Boulder, CO, USA
[3] Electrical and Computer Engineering, BYU Center for Remote Sensing, Brigham Young University, Provo, UT, USA

Describing Arctic Snow and Ice with a Small Ka-Band Radar

D. KRAMER[1,2], J. MELOCHE[1,2], A. LANGLOIS[1,2], A. ROYER[1,2], P. CLICHE[1], AND D. MCLENNAN[3]

ABSTRACT

A 24 GHz Frequency Modulated Continuous Wave (FMCW) radar is used to characterize Arctic snow and ice. This radar type has been used in mid-latitudes to describe lake ice profiles and the snowpack in mountainous regions for avalanche forecasting. As the Arctic snow is mostly wind packed, the algorithm is developed further to distinguish between the major arctic snow types, wind slabs and depth hoar. Additional development will focus on ice lens detection, as these occur more and more often due to increased rain-on-snow (ROS) events. Experiments have been conducted in Finland and Canada (Quebec, Northwest Territories, and Nunavut). A first analysis shows good results, but the radar seems very sensitive to humidity in the snow pack and has major problems if liquid water or brine are present. The accuracy is with the cm-range, but the maximum depth in ice is limited to ca. 100 cm (more in snow). The size of the radar is small enough to mount it on a remotely piloted aircraft system (RPAS), which is planned for the next campaign.

[1] Université de Sherbrooke, Sherbrooke, QC, Canada
[2] Centre d'études Nordiques, Québec City, QC, Canada
[3] POLAR Knowledge Canada, Ottawa, ON, Canada

Using Current SAR Satellite Missions to Support Future Snow Satellite Radar Missions

BENOIT MONTPETIT[1], JOSHUA KING[2], CHRIS DERKSEN[2], ANNA WENDLEDER[3], AND PAUL SIQUEIRA[4]

ABSTRACT

Landscape scale monitoring of snow and ice conditions is of great importance for wildlife habitat and ecosystem monitoring, species at risk and protected areas management and to support northern communities. Current Earth-observation missions do not provide the information needed to monitor snow mass evolution due to spatial and/or temporal scales or sensor properties such wavelengths not being sensitive to snowpack properties or sensors being dependent on solar radiation. To address this issue, Environment and Climate Change Canada (ECCC) and the Canadian Space Agency (CSA) in collaboration with many international partners have designed a dual-frequency Ku-band (17.2 and 13.5 GHz) radar mission concept at a 250 m resolution to monitor terrestrial snow mass globally. To support the mission concept, an intensive airborne, satellite and *in situ* campaign was conducted at the Trail Valley Creek (TVC) research basin, Northwest Territories, Canada. Radarsat-2 and TerraSAR-X data was acquired at least twice a month between September 2018 and April 2019 over the TVC study area to monitor changes in soil and vegetation properties. Steven's Hydra Probes were deployed at 6 different sites to measure soil temperature, moisture and electric conductivity at 4 different depths continuously over the winter season. Snow properties were also measured in December 2018, January and March 2019 around the soil stations. This unique dataset enables us to link observed signal variations of the current SAR satellite sensors to measured snow/soil properties. Preliminary results show the use of the current satellite SAR missions to support a future Ku-band radar mission to monitor terrestrial snowpack properties by decoupling soil, vegetation and soil surface roughness changes from snowpack properties.

[1] Environmental and Climate Change Canada, Landscape Science and Technology Division, Ottawa, ON, Canada

[2] Environment and Climate Change Canada, Climate Research Division, Toronto, ON, Canada

[3] German Remote Sensing Data Center (DFD), German Aerospace Center (DLR), Köln, Germany

[4] Department of Electrical and Computer Engineering, University of Massachusetts, Amherst, MA, USA

Preliminary Analysis of Ku-Band Radar Measurements over the Trail Valley Creek Region of the Canadian Northwest Territories

PAUL SIQUEIRA[1], MAX ADAM[1], CASEY WOLSEIFFER[1], JOSHUA KING[2], AND CHRIS DERKSEN[2]

ABSTRACT

During the 2018 – 2019 winter season, the University of Massachusetts and Environment and Climate Change Canada teamed up to test an airborne Ku-band interferometric SAR over the Trail Valley Creek region located in the Canadian Northwest Territories. Data were collected over three periods (November, January and March) in a grid pattern over a 60 km region, with at least three flights per period. These data were collected concurrently with ground validation and additional satellite data in order to create a suite of observations that can be used for inferring snow properties, as described in a companion paper (King *et al.*, 2019).

In this paper, a description of the UMass Ku-band SAR is given, along with intermediate analyses that have been performed on collected SAR data. These analyses include spatial and temporal characteristics of the data, and the efforts that have been employed thus far in processing the data set.

[1] Microwave Remote Sensing Laboratory, University of Massachusetts, Amherst, MA, USA
[2] Environment and Climate Change Canada, Climate Research Division, Toronto, ON, Canada

Retrieval of Snow Water Equivalent using Combinaed Microwave Active and Passive Observations

JIYUE ZHU[1], LEUNG TSANG[1], DO-HYUK "DK" KANG[2,3], AND EDWARD KIM[2]

ABSTRACT

Recently, a volume scattering approach for retrieving snow water equivalent (SWE) has been applied to three sets of airborne SnowSAR data (including 2011 and 2012 campaigns in Finland; 2013 campaign in Canada). It has achieved root-mean-square error (RSME) below 30 mm of SWE and correlation coefficients above ~0.64. In this paper, we apply the original method by including three more channels of data, the third Ku-band at ~13 GHz and radiometer observations at Ku- (~19 GHz) and Ka- (~37 GHz) bands, to form an active and passive combined method for the SWE retrieval. Introducing low Ku-band helps to alleviate the background scattering effects which is coincident with the snow mission concept of the Canadian Space Agency (CSA). Brightness temperatures (Tb) from radiometer observations are utilized to estimate the priori scattering albedo (effective grain size, one of two unknowns in the retrieval) of snow for locating the best snow parameters. The proposed combined algorithm is validated against Finland ESA NoSREx data 2009-2013 and NASA SnowEx 2017 datasets. The SnowEx 2017 winter campaign deployed both the airborne SnowSAR and the ground-based a scatterometer from the University of Waterloo to acquire radar measurements at X- and Ku-band. The Tb data applied are extracted from AMSR2 satellite observations. The NoSREx datasets include both ground-based radar measurements of X- and dual Ku- bands and radiometer observations of Ku- and Ka-bands. The RSME of the retrieval performance are below 20 mm for thin snow and less than 10% of total SWE for thick snow. The correlation coefficients are above 0.82.

[1] Department of Electrical Engineering and Computer Science, University of Michigan, Ann Arbor, MI, USA
[2] NASA Goddard Space Flight Center, Greenbelt, MD, USA
[3] ESSIC, University of Maryland, College Park, MD, USA

Machine Learning-Based Prediction of C-Band Synthetic Aperture RADAR (SAR) Backscatter over Snow-Covered Terrain

JONGMIN PARK[1] AND BARTON A. FORMAN

ABSTRACT

Snow cover and snow mass exert a significant influence on the Earth's water and energy budgets as well as on climate variability across regional and continental scales. Satellite-based observations have the advantage of capturing the spatio-temporal dynamics of snowpack information as the snow cover extent changes or snowpack deepens and/or ripens. Space-based synthetic aperture radar (SAR) backscatter observations are an attractive means of estimating snowpack information based on the variations of snow dielectric. SAR imagery also has advantages in providing multi-polarization observations at a relatively fine spatial resolution.

This study utilized support vector machine (SVM) regression to predict C-band SAR backscatter over snow-covered terrain in Western Colorado. Training targets included the co-polarized ($\sigma\sigma VVVV$) and cross- polarized ($\sigma\sigma VVVV$) backscatter coefficients as well as the ratio of those two backscatter coefficients. Inputs to the SVM were derived from the NASA Land Information System (LIS) using the NOAH-Multiparameterization (NOAH-MP) land surface model with Modern-Era Retrospective analysis for Research and Application, Version 2 (MERRA2) meteorological boundary conditions. Training periods were selected as September 2015 to August 2018 excluding the validation period selected as September 2016 to August 2017. This study particularly focuses on the influence of training period length on prediction accuracy in conjunction with the effects of data sparsity on SVM efficacy. The results highlight the strengths and weakness of machine learning in the estimation of C-band SAR backscatter over snow-covered land. A discussion on the future use of C-band SAR machine learning within an ensemble-based data assimilation framework is also presented.

[1] Department of Civil and Environmental Engineering, University of Maryland, College Park, MD, USA

Winter 2018-19 Observations with Wideband Autocorrelation Radiometry

ROGER DE ROO[1] AND MOHAMMAD MOUSAVI[2]

ABSTRACT

Wideband Autocorrelation Radiometry (WiBAR) is a technique for passively measuring the microwave propagation time of low loss layers. Radiobrightness from below the layer propagates upward toward the sensor through the layer after transiting first the lower and then the upper interfaces. At the same time, some of the radiation reflects from the upper interface, then the lower interface, before transiting the upper interface towards the sensor. This delayed ray is an attenuated and delayed copy of the direct ray, leading to a local maximum in the autocorrelation function of the received waveform. The time lag at which this maximum occurs is the round-trip propagation time of the layer. To resolve short time lags, on the order of nanoseconds, large bandwidths, on the order of gigahertz, are needed.

This technique has application for measuring the low loss layers of snow pack and lake ice. In the winter of 2018-19, we deployed two independent WiBAR instruments specifically to observe the passive lag signature of snow on the ground, one operating roughly in L-band and one in S-band. In early March 2019, a snowpack up to 64 cm had developed, which is deep enough for the WiBAR observable to be detected by each instrument. We have recently downloaded the data, and see some first hints of the signal, albeit contaminated with considerable radio frequency interference. We will report on our findings.

[1] Climate and Space Science and Engineering, University of Michigan, Ann Arbor, MI, USA
[2] Electrical Engineering and Computer Science, University of Michigan, Ann Arbor, MI, USA

Characterizing Snow Water Equivalent from Ground-Based Observations of GPS Vertical Displacement and Model-Based Hydrologic Loading Estimates

GAOHONG YIN[1], BARTON A. FORMAN[1], BRYANT D. LOOMIS[2], AND SCOTT B. LUTHCKE[2]

ABSTRACT

Winter snowpack is the most variable component of terrestrial water storage (TWS) in the mountainous regions of the Western United States. Snow water equivalent (SWE) is the most important snow characteristic with regards to snow mass and water resource forecasting. The study used ground-based GPS observations of vertical displacement to estimate SWE across snow-dominated regions of the Western United States. The SWE derived from ground-based GPS captured snow mass variations at finer spatial resolutions relative to remote sensing-based SWE retrievals. After accounting for the effects of non-hydrologic loadings on GPS-based vertical displacement, the remaining variations are predominately driven by hydrologic processes, most notably seasonal snow accumulation and ablation.

A "synthetic" experiment was first used to model SWE and soil moisture from the NASA Catchment Land Surface Model in order to evaluate the accuracy of the inversion method as well as to compute the fraction of SWE (normalized by TWS) in the study area. Afterwards, a "real-world" experiment was conducted using ground-based GPS observations from the Plate Boundary Observatory network. The inverted TWS was then compared against TWS retrievals derived from the Gravity Recovery and Climate Experiment (GRACE) mission. Retrieved SWE was validated using SWE observations from the ground- based Snow Telemetry (SNOTEL) network. Preliminary results show that over half of the stations provide a temporal correlation coefficient of $R < -0.7$ between GPS-based vertical displacement and SNOTEL-based SWE, which highlights the dominant effect of snow on the subsequent vertical displacement variation.

[1] Department of Civil and Environmental Engineering, University of Maryland, College Park, MD, USA
[2] Geodesy and Geophysics Laboratory, NASA Goddard Space Flight Center, Greenbelt, MD, USA

UAV LiDAR for Measuring Snow Interception in Forests

COB STAINES[1] AND JOHN W. POMEROY[1]

EXTENDED ABSTRACT

Uncrewed aerial vehicle (UAV) LiDAR presents opportunities to advance knowledge of snow interception by examining the relationships between subcanopy snow accumulation and needleleaf forest canopy structure. A robust understanding of how snow interception varies with canopy structural metrics will aid in estimating snowmelt water availability for streamflow. This could improve drought and flood forecasting, understanding of effects of land disturbance and timber harvesting on hydrology, and anticipation of hydrological responses to future weather patterns in a changing climate (see Varhola et al., 2010 for a review).

Decades of study and field trials have shown that measuring intercepted snow directly in needleleaf canopies is logistically prohibitive at scales beyond that of individual trees (Friesen et al., 2015). Interception models are developed and employed to estimate rates of interception processes but often differ in assumptions of relationships between interception processes and canopy structure metrics (e.g. Hedstrom and Pomeroy, 1998; Moeser et al., 2015; Roth and Nolin, 2019). Further validation is needed to inform which assumptions are most appropriate for a given forest under a given set of environmental conditions, and at which scales.

Snow mass budgeting is a method used to estimate missing terms in the snow mass balance. This can be used to estimate changes in intercepted load over a time interval as the difference between cumulative precipitation above and accumulation below the canopy, corrected for other system fluxes such as sublimation and wind redistribution of snow (Pomeroy and Schmidt, 1993; Lundberg et al., 1998). The spatial resolution and extent of interception estimates from the mass budgeting method depends on the resolution and extent of snow accumulation measurements.

Repeated airborne LiDAR scans have been used to calculate snow depth at horizontal resolutions on the order of 1m and over the scale of basins (Deems et al., 2013). Airborne LiDAR has also been used to characterize the structure of forest canopies with applications to both snowpack energy and mass balances (Broxton et al., 2015; Moeser et al., 2015; Mazzotti et al., 2019). This technology has encountered challenges in dense vegetation due to reduced canopy penetration by the LiDAR beam (Harpold et al., 2014; Broxton et al., 2015; Zheng et al., 2015). High costs have limited the temporal resolution of airborne LiDAR scans in most applications. UAV LiDAR shows promises in overcoming these challenges with reduced overhead and operational costs, allowing for increased temporal resolution of samples (Harder et al., 2016), and increased spatial resolution due to smaller footprints, tighter point spacing, and variable-perspective scans (Popescu et al., 2011; Wallace et al., 2012).

[1] Centre for Hydrology, University of Saskatchewan, Saskatoon, SK, Canada

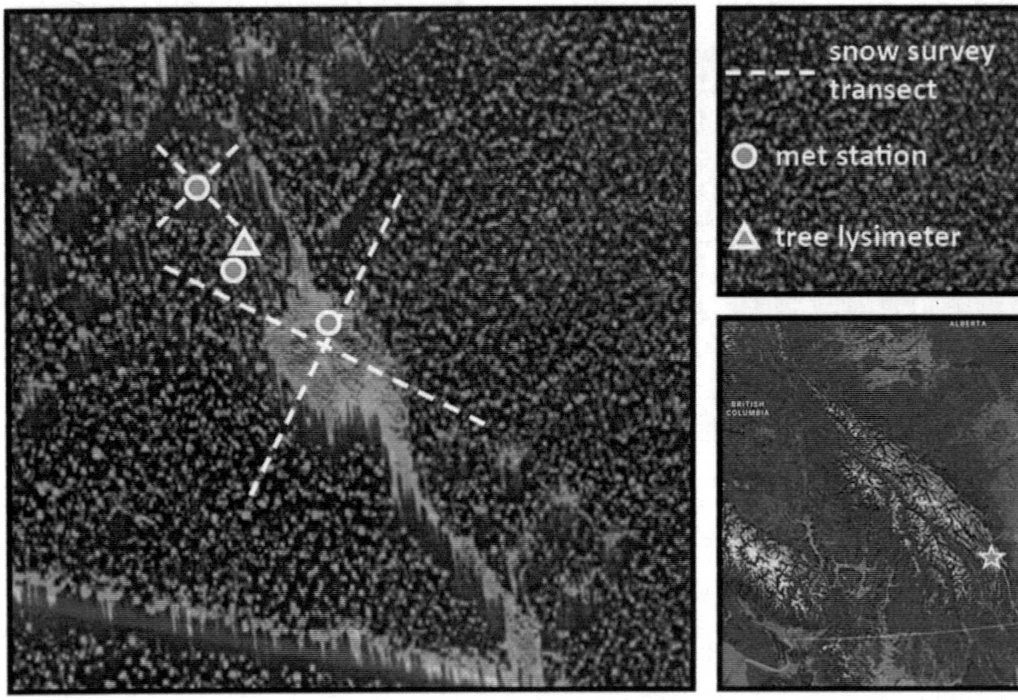

Figure 1. Layout of field site in the Canadian Rockies depicting the location of snow survey transects, three meteorological stations, and the tree lysimeter.

The purpose of this research is to quantify the spatial variation of intercepted snow with canopy structure at scales from trees to forest stands. Six UAV surveys were conducted within the 2018-2019 winter season over an instrumented forest site at Marmot Creek Research Basin on the eastern slopes of the Canadian Rockies (Figure 1). A DJI M-600 hexcopter UAV mounted with a RIEGL miniVUX-1UAV LiDAR sensor and co-registered Sony a6000 camera collected point cloud and RGB data over a 500 m-by-500 m area from tiered flight paths at altitudes ranging from 40 m-120 m above ground level (see Figure 2). Each UAV survey was accompanied by ground-based snow surveys with georeferenced snow depth and density observations, and upward-facing hemispherical photography. Meteorological data was collected at three stations within the survey area: one below the forest canopy and two in nearby clearings (Figure 1). Wind speed and direction, air temperature and humidity, net radiation and snow depth time series were collected below the canopy and in clearings. Cumulative precipitation was measured in the larger clearing by a Geonor T-200B precipitation gauge. A hanging tree lysimeter adjacent to the forested station provided the mass of intercepted snow in a single tree through the winter season.

Flight paths at multiple elevations were implemented to increase the spatial resolution of subcanopy returns, driven by an increase in the number of possible scanner perspectives of a given ground point. Preliminary results show median forest point densities from the subset of scans at 120m above ground level near 75 points/m2 (Figure 3), and the ability to resolve individual branches within the upper canopy.

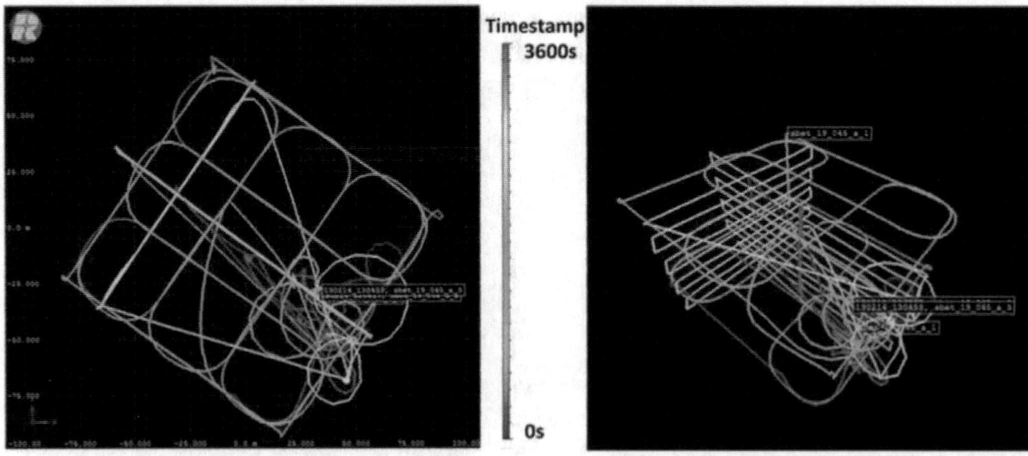

Figure 2. Ice cover (blue) in the middle and lower Chesapeake Bay in Early February 1977.

Further results are expected to inform an understanding of the role of canopy structure in snow interception processes and facilitate the validation and continued development snow interception models. Interception will be estimated from this data using the mass budgeting method. LiDAR derived snow depth models will be validated with manual measurements and converted to snow water equivalent (SWE) using depth-density relationships from concurrent snow survey observations. The change in SWE will be calculated as the difference between successive SWE maps. Canopy structure metrics (e.g. leaf area index and canopy closure) will be calculated from snow-off LiDAR point clouds and validated using similar metrics derived from hemispherical photography. The spatial variation of interception estimates with canopy metrics will be assessed. Analysis will be conducted for interception estimates and canopy structural metrics aggregated to different scales to assess the scale dependence of this method.

REFERENCES

Broxton PD, Harpold AA, Biederman JA, Troch PA, Molotch NP, Brooks PD. 2015. Quantifying the effects of vegetation structure on snow accumulation and ablation in mixed-conifer forests. *Ecohydrology*, **8**(6): 1073–1094. https://doi.org/10.1002/eco.1565

Deems JS, Painter TH, Finnegan DC. 2013. Lidar measurement of snow depth: A review. *Journal of Glaciology*, **59**(215): 467–479. https://doi.org/10.3189/2013JoG12J154

Friesen J, Lundquist J, Van Stan JT. 2015. Evolution of forest precipitation water storage measurement methods. *Hydrological Processes*, **29**(11): 2504–2520. https://doi.org/10.1002/hyp.10376

Harder P, Schirmer M, Pomeroy JW, Helgason W. 2016. Accuracy of snow depth estimation in mountain and prairie environments by an unmanned aerial vehicle. *The Cryosphere*, **10**(6): 2559–2571. https://doi.org/10.5194/tc-10-2559-2016

Harpold AA, Biederman JA, Condon K, Merino M, Korgaonkar Y, Nan T, Sloat LL, Ross M, Brooks PD. 2014. Changes in snow accumulation and ablation following the Las Conchas Forest Fire, New Mexico, USA. *Ecohydrology*, **7**(2): 440–452. https://doi.org/10.1002/eco.1363

Hedstrom NR, Pomeroy JW. 1998. Measurements and modeling of snow interception in the boreal forest. *Hydrological Processes*, **12**(10): 1611–1625. https://doi.org/10.1002/(SICI)1099-1085(199808/09)12:10/11<1611::AID-HYP684>3.0.CO;2-4

Lundberg A, Calder I, Harding R. 1998. Evaporation of intercepted snow: measurement and modelling. *Journal of Hydrology*, **206**(3–4): 151–163. https://doi.org/10.1016/S0022-1694(97)00016-4

Mazzotti G, Currier WR, Deems JS, Pflug JM, Lundquist JD, Jonas T. 2019. Revisiting snow cover variability and canopy structure within forest stands: Insights from airborne lidar data. *Water Resources Research*, **55**(7): 6198-6216. https://doi.org/10.1029/2019WR024898

Moeser D, Stähli M, Jonas T. 2015. Improved snow interception modeling using canopy parameters derived from airborne LiDAR data. *Water Resources Research*, **51**(7), 5041–5059. https://doi.org/10.1002/2014WR016724

Pomeroy JW, Schmidt RA. 1993. The use of fractal geometry in modelling intercepted snow accumulation and sublimation. *50th Annual Eastern Snow Conference*, 231–239. Quebec City.

Popescu SC, Zhao K, Neuenschwander A, Lin C. 2011. Satellite lidar vs. small footprint airborne lidar: Comparing the accuracy of aboveground biomass estimates and forest structure metrics at footprint level. *Remote Sensing of Environment*, **115**(11): 2786–2797. https://doi.org/10.1016/j.rse.2011.01.026

Roth TR, Nolin AW. 2019. Characterizing maritime snow canopy interception in forested mountains. *Water Resources Research*, **55**(6): 4564-4581. https://doi.org/10.1029/2018WR024089

Varhola A, Coops NC, Weiler M, Moore RD. 2010. Forest canopy effects on snow accumulation and ablation: An integrative review of empirical results. *Journal of Hydrology*, **392**(3–4): 219–233. https://doi.org/10.1016/j.jhydrol.2010.08.009

Wallace L, Lucieer A, Watson C, Turner D. 2012. Development of a UAV-LiDAR system with application to forest inventory. *Remote Sensing*, **4**(6): 1519–1543. https://doi.org/10.3390/rs4061519

Zheng Z, Kirchner PB, Bales RC. 2015. Orographic and vegetation effects on snow accumulation in the southern Sierra Nevada: A statistical summary from LiDAR data. *The Cryosphere Discussions*, **9**(4): 4377–4405. https://doi.org/10.5194/tcd-9-4377-2015

High Resolution Shallow Snowpack Snow Depth Variability from Unmanned Aerial Systems (UAS) Mounted LiDAR Observations

ADAM HUNSAKER[1], JENNIFER M. JACOBS[1], MICHAEL PALACE[1], FRANKIE SULLIVAN[1], AND RONNY SCHROEDER[1]

ABSTRACT

In order to downscale coarse global satellite observations of snow depth and snow water equivalent, a deeper understanding of how physical drivers influence snow spatial variability is needed. UAS platforms offer the potential to make high spatiotemporal resolution snow depth observations at small watershed scales with a high degree of accuracy at spatial scales unattainable with satellite observations. During the winter 2018-2019, UAS LiDAR and RGB imagery observations were made in Durham, NH over approximately 35 acres including large open and forested areas. *In situ* measurements were collected to assess the accuracy of the UAS derived snow depth maps. Here we provide preliminary results about snow depth variability from multiple shallow (~ 10 cm) snowpacks.

[1] University of New Hampshire, Durham, NH, USA

Improvements to the Interactive Multisensor Snow and Ice Mapping System (IMS) and Advantages of IMS over Automated Snow Cover Detection Algorithms

J. EDWARDS-OPPERMAN[1], M. LOWE[1], D. MCCORMICK[1], J. WOODS[1], AND K. BERBERICH[1]

ABSTRACT

Accurate initial conditions of snow and ice cover are essential for operational numerical weather prediction (NWP) models. Analysts at the US National Ice Center use the Interactive Multisensor Snow and Ice Mapping System (IMS) software platform to provide twice daily (18Z and 0Z) analyses of snow and ice cover in the Northern Hemisphere. These data are processed at 1-, 4-, and 24-km resolution and are used in many operational weather models as well as in various research efforts.

The IMS has several advantages over automated snow cover detection algorithms. Analysts are not limited by cloud cover (as are many automated satellite-based snow detection algorithms) and can make judgments about snow cover based on a variety of *in situ* data including weather stations, webcams, and public reports. Additionally, many automated algorithms have trouble picking up snow cover in forests due to obstruction by the canopy. Analysts can also use *in situ* observations, weather models, radar, and other meteorological data to provide short-term forecasts ('nowcasts') of snow cover, thus providing accurate initial conditions at a consistent time for NWP models.

Recent data additions to the IMS platform, including visible satellite imagery from GOES 16 and 17, passive microwave data from the Advanced Microwave Scanning Radiometer 2, more timely synthetic aperture radar imagery from the Sentinel-1 and Radarsat satellites, and more frequent updates of WSR-88D radar data, have enhanced analyst ability to determine areas of snow and ice cover. These new data sources for the IMS have improved the accuracy of initial snow and ice conditions input into NWP models.

[1] U.S. National Ice Center, Washington, DC, USA

Duration of Snow Cover in the Western U.S. Measured using MODIS and VIIRS Cloud-Gap-Filled Snow Cover Products

DOROTHY K. HALL[1,2], GEORGE RIGGS[2,3], AND NICOLO E. DIGIROLAMO[2,3]

ABSTRACT

A great deal of recent work has shown that the duration of snow cover in North America has been decreasing since the satellite snow-cover record began in the late 1960s. This reduction in snow-cover duration, notably resulting in increasingly earlier snowmelt, has been particularly evident in coastal ranges in western North America such as the Pacific Northwest. In this work, we use the new MODerate-resolution Imaging Spectroradiometer (MODIS) and Visible Infrared Imaging Radiometer Suite (VIIRS) cloud-gap-filled daily snow-cover products to study changes in the duration of snow cover in selected mountain ranges along the western coast of the United States such as the Cascade Range. We also investigate changes in the continental interior such as the Wasatch, Wind River, and Sierra Nevada mountain ranges over an 18-year period, from the winter of 2000-2001 through the winter of 2017-2018.

[1] Earth System Science Interdisciplinary Center (ESSIC), University of Maryland, College Park, MD, USA
[2] NASA Goddard Space Flight Center, Greenbelt, MD, USA
[3] Science Systems and Applications, Inc., Greenbelt, MD, USA

VIIRS and MODIS Cloud-Gap-Filled Snow Cover Products in New Data Collections

GEORGE RIGGS[1,2] AND DOROTHY K. HALL[3,2]

ABSTRACT

New NASA VIIRS and MODIS daily cloud-gap-filled (CGF) snow cover data products released in the new MODIS and VIIRS data collections are described and exhibited. The NASA VIIRS daily CGF snow cover data product (VNP10A1F) is released in the Land Science Investigator-led Processing System (LSIPS) V2 data processing collection beginning about June 2019. The MODIS daily CGF snow cover product for both Terra (MOD10A1F) and Aqua (MYD10A1F) are produced in the MODIS Adaptive Processing System (MODAPS) C6.1 data processing collection beginning about May 2019. These data products will be available from the National Snow and Ice Data Center (NSIDC) NASA Data Active Archive Center (DAAC). Data product formats and contents are described to enable a user to better understand the products. Daily CGF snow cover extent (SCE) maps from the products are shown in comparison to the standard SCE maps and seasonal records of snow cover generated from the CGF products for select regions in the western USA.

[1] Science Systems and Applications, Inc., Greenbelt, MD, USA
[2] NASA Goddard Space Flight Center, Greenbelt, MD, USA
[3] Earth System Science Interdisciplinary Center (ESSIC), University of Maryland, College Park, MD, USA

Spectral Reflectance Signatures of Compacted Snow Surfaces

S. SHOOP[1], B. ELDER[1], AND D. PEROVICH[2]

ABSTRACT

Standoff assessment of snow physical properties is useful for making estimates of snow conditions that would be suitable for vehicle or aircraft operations prior to *in situ* strength measurements. Satellite or UAV is one method to obtain data, but to compare on-site with satellite data, a handheld field spectrometer was also used to collect data at a range of site conditions. The full spectral reflectance waveform was collected using an ASD Field-Spec4 Hi-Res spectroradiometer. Reflectance measurements from 350-2500nm with 3–8 nm of spectral resolution were collected at target locations including packed, groomed and natural snow, in addition to asphalt and ice surfaces in Montana and northern Michigan. At the same time, a suite of measurements was also collected to characterize the snow physical and mechanical properties, along with concurrent satellite imagery from the WorldView3 satellite. The aim was to determine if and how the snow physical properties could be inferred from these stand-off (non-tactile) measurement techniques. The spectra were compared to strength measurements taken at each location. While many show no correlation, two near surface measurements show promising trends. In additional, the spectral signatures characteristic of each type of snow surfaces is distinguishable and will be more fully explored.

[1] U.S. Army Corps of Engineers Cold Regions Research and Engineering Laboratory, Hanover, NH, USA
[2] Dartmouth College, Hanover, NH, USA

Snow Estimation in Complex Terrain using the NASA Land Information System

JAWAIRIA A. AHMAD[1], BARTON A. FORMAN[1], SUJAY KUMAR[2], AND EDWARD BAIR[3]

ABSTRACT

Snow estimation in complex terrain at continental scales has remained as yet, an unsolved problem due to the high spatial variability in elevation, relatively coarse resolution of available satellite data, presence of clouds, and a general lack of ground observations for validation and evaluation purposes. In this study, a machine learning integrated data assimilation framework is utilized to estimate snow in high mountain Asia. The NASA Land Information System (LIS) is the software framework used here to simulate the hydrologic cycle and to assimilate brightness temperature spectral difference observations using an ensemble Kalman filter. Trained support vector machines act as the observation operator within the assimilation framework and map the LIS simulated geophysical states into brightness temperature spectral difference space. Snow estimates (with and without assimilation) are compared to ground-based observations for performance evaluation. Recently acquired *in situ* snow depth measurements are translated to snow water equivalent values using downscaled meteorological forcings and the SNOWPACK model. The assimilation framework exhibits potential in improving the land surface model-based snow estimates. However, machine learning pitfalls such as controllability and under-determined systems do exist at certain locations in time and highlight some of the challenges of utilizing machine learning algorithms as observation operators within a data assimilation framework.

[1] Department of Civil and Environmental Engineering, University of Maryland, College Park, MD, USA
[2] Hydrological Sciences Laboratory, NASA Goddard Space Flight Center, Greenbelt, MD, USA
[3] Earth Research Institute, University of California, Santa Barbara, CA, USA

Development of a Numerical Roof Snow Load Model

STEVEN HALL[1] AND JUSTIN FERRARO[1]

ABSTRACT

Snow loads on roofs can be calculated with a numerical model that combines velocities over the roofs with historical climate data. A model has been created that obtains the velocities from Computational Fluid Dynamics (CFD) simulations and iterates over several years of climate data to produce estimated snow loads over the roofs. This model considers the effects of snowfall, rainfall, melting, freezing, and drifting over the full course of each winter season. Several advantages and challenges are discussed regarding the use of CFD instead of physical experiments. Finally, several areas for future development and research are discussed.

[1] Gradient Wind Engineering Inc., Ottawa, ON, Canada

Brightness Temperatures of Snowpack from Microwave Radiative Transfer Models (RTM) by using Two Separate Drivers: 1) Snow Physics Model Outputs, and 2) *in situ* Snowpit Stratigraphy

DO HUK "DK" KANG[1,2], SHURUN TAN[3], AND EDWARD J. KIM[2]

ABSTRACT

This paper evaluates different brightness temperature (Tb) from three microwave radiative transfer models (RTM) of snowpack simulated by 1) snow physics model outputs, and 2) *in situ* snow stratigraphy observations. A set of evaluations is conducted by simulating the RTM (HUT among three RTMs) with the output of a snow physics model driven by actual weather forcing in a coupled simulation. Outputs of this coupled model include snowpack physical properties and Tbs. Another part of Tb simulation is also included with RTMs driven by *in situ* snowpit stratigraphy observations. The snow physics outputs from the coupled case are compared against *in situ* snow stratigraphy measurements from the European Space Agency Nordic Snow Radar Experiment (NoSREx) 2009-2012. And, both RTM and *in situ* driven Tb simulations are compared against ground-based microwave observations also at NoSREx. The paper suggests a temporarily divisional approach to interpret microwave Tb to relate snow conditions along the snow year, e.g. Phase 1~4, from accumulation to melt. For three consecutive years, 2009-12, the *in situ* driven Tbs have 21.0 K Root Mean Squared Error (RMSE), while the coupled simulations have 24.7 K RMSE. Particularly, the water year 2011 is divided into 4 Phases. In the Phase 3 in 2011, 12.2 K and 6.3 K RMSEs are achieved from *in situ* and coupled cases, respectively. Such an RMSE improvement (6.3) from the coupled case in the Phase 3 is made after isolating the dry snow period and excluding diurnal melting snow conditions.

[1] Earth Systems Science Interdisciplinary Center, University of Maryland, College Park, MD, USA
[2] NASA Goddard Space Flight Center, Greenbelt, MD, USA
[3] Zhejiang University / University of Illinois at Urbana-Champaign Institute, Haining, Zhejiang, China

The Application of SnowModel to Vehicle Mobility in Winter

TED LETCHER[1], MICHELLE MICHAELS[1], AND JULIE PARNO[1]

ABSTRACT

Vehicle mobility in snow is of particular interest to the U.S. Army. We applied SnowModel, a spatially distributed, physically-based snow evolution modeling system, to characterize snow in areas where military vehicles are tested in a greater effort to help determine vehicle mobility limitations in snow. We used SnowModel to simulate a full winter season over small domains in Wyoming, Michigan, and Vermont. Meteorological forcing for the model is generated from weather station data archived in the Integrated Surface Database and the Global Historical Climatology Network. In each domain, the model is run on a 10-meter grid and simulates snow accumulation and ablation and captures seasonal and spatial snowpack variability. Additional processes represented in the model include snow densification, blowing-snow redistribution and sublimation, interception, unloading, and sublimation within forest canopies, and snowpack ripening. To increase efficiency, we parallelized the SnowModel code and implemented it on a high-performance computing system, resulting in as much as a 95% decrease in model run time for domains with a large number of weather stations. Preliminary model results will be presented.

Keywords: GHCN, snow depth, snow water equivalent, snow liquid ratio, snow class

INTRODUCTION

Vehicle mobility in snow is of particular interest to the U.S. Army. Snow strength as it relates to vehicle mobility is particularly complex, especially over transient snowpacks that are subject to episodic freeze/thaw cycles throughout the winter season (Birkeland *et al.*, 1995; Pielmeier and Schneebeli, 2003). While snow depth and snow density are critical snow characteristics that impact vehicle traction and floatation (e.g., Blaisdell *et al.*, 1990), snowpack stratigraphy, liquid water content, presence of ice-layers and depth hoar are also important (e.g., Shoop *et al.*, 2006; Pytka, 2010). Snow observations are often far and few between and are typically limited to snow depth, making assessing snow strength from these observations a challenge. To better predict snow cover impacts on vehicle mobility at fine spatial scales, a numerical model that can simulate the complex physical processes that determine snow strength is an essential tool. In this study, we use SnowModel, a spatially distributed, physically-based snow evolution modeling system, to characterize snow in areas where military vehicles are tested in a greater effort to help determine vehicle mobility limitations in snow.

SnowModel (Liston and Elder, 2006a) is a widely used and well-documented distributed energy and mass balance snow model. The meteorological forcing is determined via optimal interpolation and terrain downscaling of available surface station data to the model grid (Liston and Elder, 2006b). The SnowModel system is designed to be run on very high resolution grids ($\Delta x < 100$ m) and

[1] U.S. Army Corps of Engineers Cold Regions Research and Engineering Laboratory, Hanover, NH, USA

simulates the most important processes driving the accumulation and ablation of snow including: terrain slope and effects on incoming shortwave radiation; overhead canopy and cloud cover effects on shortwave and longwave radiation; canopy interception/unloading and sublimation of snowfall; snowpack densification and ripening; and liquid refreeze within the snowpack. Finally, a blowing snow transport model (Liston and Sturm, 1998) is incorporated into the SnowModel framework and simulates the lateral movement and enhanced sublimation of snow by the wind.

To correct for typically challenging and uncertain snow precipitation observations, a "data assimilation" technique can be applied within the SnowModel framework with the aim of accounting for precipitation undercatch. The assimilation technique uses available snow depth measurements and determines a correction factor, which is then used to backfill the precipitation forcing such that it closes the snow deficit between the model and observations at assimilation time. The intrinsic assumption underpinning this technique is that precipitation undercatch is the largest contributor to model error throughout the winter season, and that other model errors are secondary in nature.

While SnowModel has a high degree of physical realism, several processes impacting the surface energy balance are either oversimplified or neglected entirely. For instance, SnowModel has a fixed dry-snow albedo, and ignores the precipitation heat flux. Additionally, because SnowModel is designed to run with surface station data, the parameterizations for cloud cover, incoming longwave radiation, and temperature/wind/precipitation lapse rates are subject to large uncertainty.

The primary goal of this study is to investigate SnowModel as a predictive tool for snow strength metrics by assessing its ability to accurately simulate snow depth and snow density at different model resolutions over three Army training sites located within the United States with distinct terrain and land cover properties. By focusing primarily on model resolution, the effects of terrain and land cover on the modeled snowpack evolution can be evaluated.

DATA AND METHODS

We perform SnowModel simulations over three domains within the continental United States (Fig. 1). The simulations are run at three grid-spacings: $\Delta x = 10$ m, $\Delta x = 100$ m, and $\Delta x = 1000$ m to assess the impact of resolution on the simulated snowpack characteristics. From west-to-east, the study domains are West Yellowstone, the Keewenaw Research Center located on Michigan's Upper Peninsula (hereby KRC), and Camp Ethan Allen Training Site in Jericho, Vermont (hereby CEATS). The 10-m terrain and land cover grids used as the model static input are presented in Figure 2, and key domain statistics are located in Table 1. The domain sizes range from ~ 600 km^2 for KRC to ~3000 km^2 for West Yellowstone, and each domain contains a wide range of land cover types including mixed and evergreen forests, grass and shrub lands, and urban centers and road corridors (Fig. 2). The CEATS domain has the most variable terrain with nearly 1220 m of elevation between the domains highest and lowest grid point and KRC has the least terrain variability with only a 183 m maximum elevation difference.

In all three domains, the meteorological forcing for SnowModel is generated from surface station data archived in the National Oceanic and Atmospheric Administration (NOAA) Integrated Surface Database (ISD: Smith *et al.*, 2011). The ISD aggregates quality controlled global hourly and synoptic (6-hourly) observations into a common ASCII format and includes standard meteorological parameters such as windspeed and direction, near surface air temperature and dewpoint, surface pressure, and accumulated precipitation. These data are aggregated into 3-hour average windspeed, direction, temperature, and dewpoint, and 3-hour accumulated precipitation for each forcing site. In each domain, only one ISD station is available (Fig. 2). Note that for the CEATS domain, the ISD station used is the surface weather station at the Burlington, VT airport, which is approximately 17 km west of the westernmost edge of the domain.

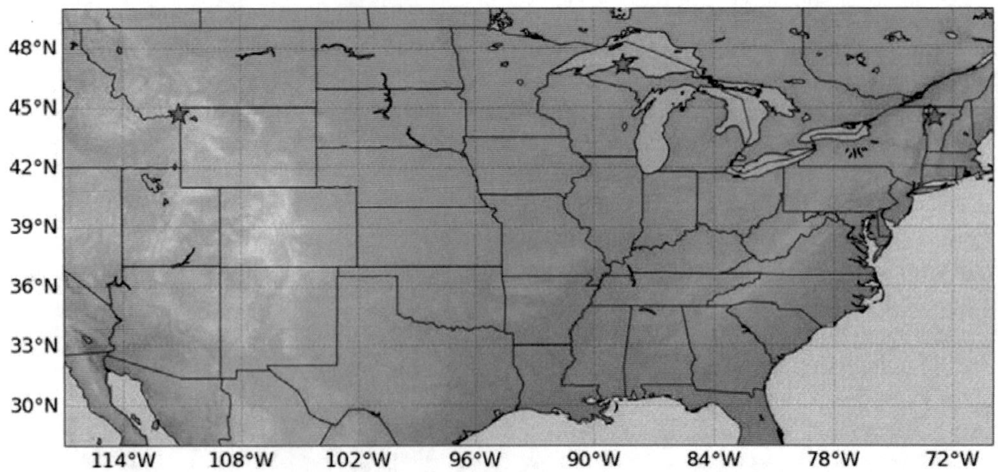

Figure 1. Map of United States with the three domain locations marked as red stars.

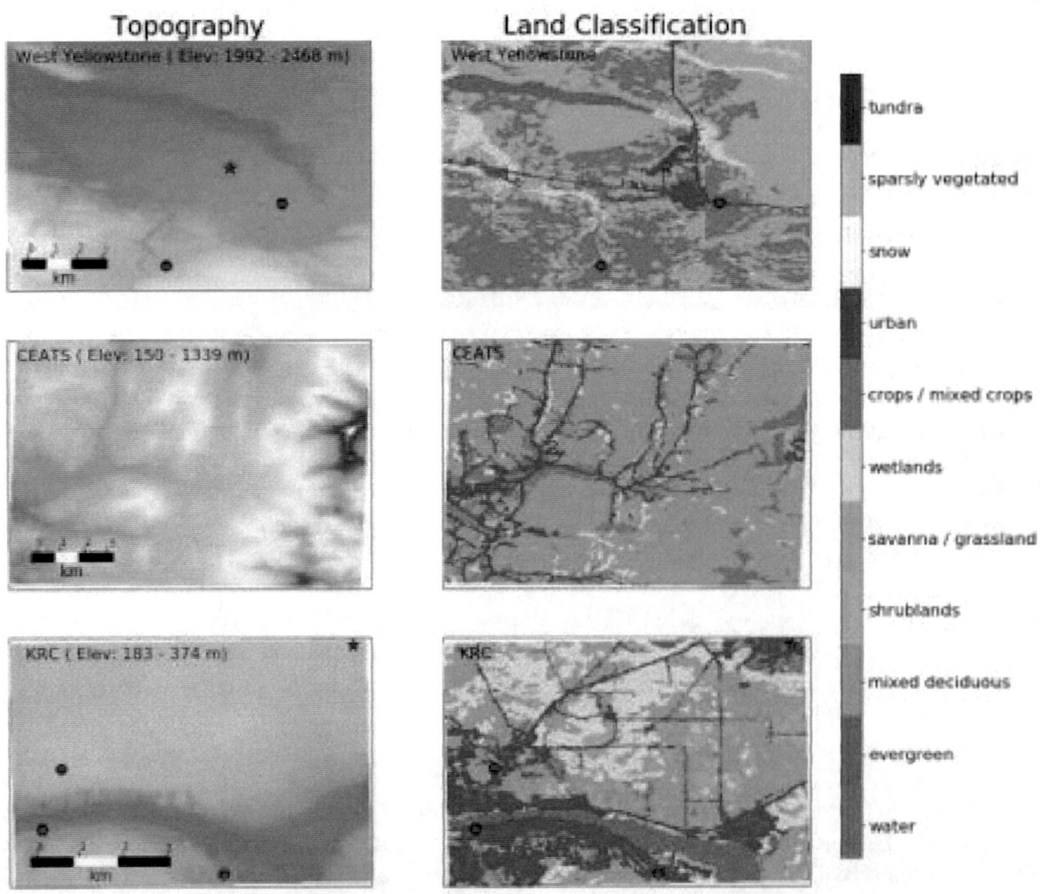

Figure 2. Detailed view of each domain. Topography is shown in the left column, and land classification is shown in the right column. ISD stations are marked as red stars, and GHCN stations are marked as blue circles. In each domain, the terrain is plotted for a 1219 m (4000 ft) range starting from the lowest domain elevation. Note that forcing station used for the Camp Ethan Allan domain is approximately 17 km west of the western domain edge and not shown in the figure.

Table 1. Summary statistics for each of the three SnowModel domains. All statistics are based on the 10-m resolution grid.

Name	Domain extent	Elevation	Forest fraction	Detailed forest
West Yellowstone	302.9 km^2	Lo: 1992 m Hi: 2468 m	53.4%	Coniferous: 53.3% Deciduous: 0.1% Mixed: 0% Short Coniferous: 0% Clear cut: 0%
KRC	59.7 km^2	Lo: 183 m Hi: 374 m	82.1%	Coniferous: 11.5% Deciduous: 57.4% Mixed: 13.4% Short Coniferous: 0% Clear cut: 0%
CLEATS	252.2 km^2	Lo: 150 m Hi: 1339 m	43.8%	Coniferous: 2.3% Deciduous: 33.7% Mixed: 7.7% Short Coniferous: 0% Clear cut: 0%

To supplement data from the ISD at the West Yellowstone and KRC domains, daily summary data from the Global Historical Climatology Network (GHCN: Peterson and Vose, 1997) is incorporated into the model forcing. These data only include daily average temperature and daily accumulated precipitation, and therefore do not affect the wind or humidity forcing.

The point station data are mapped to the gridded SnowModel domain using a Barnes optimal interpolation routine to distribute the point data across the grid. For the CEATS domain, no interpolation is applied since there is only one station in the forcing dataset. At this domain, the station data is applied uniformly across the grid.

The gridded station data is downscaled to the terrain using climatological lapse rates for temperature, moisture, and precipitation. No elevation adjustment is applied to windspeed because the physical processes that determine windspeed are much more complex and variable than the processes that determine temperature and moisture lapse rates. Minor terrain adjustments are applied to the windspeed and direction following Liston and Elder (2006b).

SnowModel Overview

Primarily, SnowModel is a mass and energy balance column model that solves the energy balance equation:

$$M = Q_s(1-\alpha) + Q_l - SH - LE - \varepsilon\sigma T_s^4 - G + L \quad (1)$$

and the mass balance equation:

$$\frac{dSWE}{dt} = P - E - R \quad (2)$$

where, in the energy balance, G is the conductive transfer of heat through the snowpack, Q_s is the downwelling shortwave radiation, α is the surface albedo, Q_l is the downwelling longwave radiation, SH and LE are the turbulent sensible and latent heat fluxes, respectively, L is the liquid refreeze heat flux, M is the melt flux, ε is the emissivity, σ is the Stefan-Boltzmann constant (5.67 x 10^{-8} W m^{-2} K^{-4}), and T_s is the snow surface temperature. In the mass balance, P, E, and R are rates of precipitation, evaporation/sublimation, and runoff, respectively. No distinction is made between frozen and liquid precipitation in the mass balance equation, since liquid is allowed to be retained in the snowpack via refreezing. The snowmelt energy flux (M) is computed as residual of the energy budget when T_s is determined to be above freezing, i.e., it is the energy required to maintain a snow surface temperature of freezing. Runoff is computed each timestep as the residual liquid (rain + snow melt) not retained within the snowpack.

Critically, a number of parameterizations are required to retrieve the energy and mass balance forcing terms from surface data. In particular, Q_s and Q_l are difficult to obtain from standard surface meteorological observations since they are strongly dependent on the 3D atmosphere. In SnowModel, a baseline Q_s is determined from latitude, time of day, elevation, slope and aspect, and day of year. It is then adjusted for cloud cover, where cloud cover is parameterized from surface temperature and moisture following Walcek (1994). Longwave radiation is parameterized from surface temperature and cloud cover following Iziomon *et al.* (2003). Finally, at grid cells where the land cover classification is vegetated, a canopy adjustment is made to Q_s and Q_l following Liston and Elder (2006a).

Precipitation observations are not intrinsically categorized as rain or snow, rather the rain/snow partitioning is performed in SnowModel using a simple temperature threshold:

$$f_s = \begin{cases} 1, & T_{air} \leq 0\ ^\circ\mathrm{C} \\ 0, & T_{air} > 0\ ^\circ\mathrm{C} \end{cases} \quad (3)$$

where f_s is the fraction of precipitation that falls as snow. At vegetated grid cells, the canopy can intercept a fraction of the snowfall where it is stored until it either sublimates or is added to the surface snowpack at a rate determined by the air temperature following Liston and Elder (2006a).

When snow transport is included in the model, the mass balance equation becomes:

$$\frac{dSWE}{dt} = P - E - R - \nabla \cdot Q_t - E_t \quad (4)$$

where $\nabla \cdot Q_t$ is the horizontal divergence of transported snow, and E_t is an additional sublimation term applied to the transported snow. $\nabla \cdot Q_t$ and E_t are computed following the parameterization described in Liston and Sturm (1998). Critically, this parameterization assumes that all snow is equally susceptible to wind transport regardless of character, and that all blowing snow is given a constant density.

Baseline Simulations

We first perform a baseline simulation for each domain using a common model configuration. These simulations are performed for a single winter season beginning in September and ending in June. The common simulations are performed with a horizontal grid-spacing of 10 m, a 3-hour time step, and with tunable parameter options as given in Table 2. These simulations are performed without data assimilation.

Table 1. Summary statistics for each of the three SnowModel domains. All statistics are based on the 10-m resolution grid.

Parameter description	Value
Shortwave radiation canopy gap fraction adjustment to canopy transmissivity	0.1
Threshold friction velocity for snow transport	0.25 m s^{-1}
Blowing snow density	300 kg m^{-3}
Dry snow albedo	0.8
Wet snow albedo for vegetated grid-cells	0.45
Wet snow albedo for non-vegetated grid-cells	0.6
Terrain curvature length scale for wind adjustment	700 m
Terrain slope weight for wind adjustment	0.58
Terrain curvature weight for wind adjustment	0.42

Experimental Configurations

The primary interest of this work is to investigate the effects of model resolution on snow depth and density as these variables have a strong impact on vehicle mobility. We chose to run the model at 100-m and 1000-m horizontal resolution over each domain. In these configurations, the 10-m terrain dataset is aggregated and averaged to match the coarser resolution grid. The 30-m land cover dataset is resampled to the coarser grid, assigning the majority land cover classification value within the coarse filter window.

Since SnowModel is a column land model, rather than a 3D Eulerian dynamic model, this is primarily an investigation of the impact that static terrain and prescribed downscaling has on snow since the underlying meteorological forcing is identical at all three resolutions. Critically, this experiment quantifies the benefits of performing high resolution simulations in data-sparse regions, and illuminates upon the relative roles of meteorology and terrain in determining snowpack evolution. Additionally, because SnowModel does not parameterize sub-grid partitioning of land or snow cover, model accuracy is expected to degrade with decreasing resolution as sub-grid effects become increasingly important.

Model Evaluation

The simulated snowpack is evaluated against any available snow depth and SWE *in situ* data within each domain. In the CEATS and KRC domains, only snow depth data is available and, accordingly, in these domains, model snow density cannot be assessed.

In West Yellowstone, two automated SNOw TELemetry (SNOTEL) stations are available, providing daily average snow depth and SWE data, thus allowing for an evaluation of model snow density. In addition to SNOTEL observations, high density snow depth observations collected at the training site during mid-winter are used to evaluate the model's ability to reproduce fine-scale snow depth variability at West Yellowstone.

At CEATS, there are five automated GHCN stations with snow depth data available. Of these, two are located in Underhill, VT, in the north central portion of the domain, two are located in Jericho, VT, on the western edge of the domain, and one is atop Mt. Mansfield at 1,204 m in elevation. The Underhill and Jericho stations are all at relatively lower elevations, ranging from 195 to 285 m, allowing us to further investigate the effect of terrain on modeled snow outputs.

At KRC, snow depth data are available at three GHCN stations. Two are located in Houghton, MI – one in the northwest, and one in the southeastern region of the domain. The third station in Quincy Hill, MI is located in the southwestern corridor of the domain. While all three stations sit at similar low elevations, the most northwestern station is separated from the others by a sizable body of water, Portage Lake. This allows for analysis of how well the model handles a variety of land cover types in a relatively small area, including water, which is unique among the testing sites.

RESULTS

While SnowModel was able to reproduce the winter accumulation and spring ablation of the snowpack with reasonable fidelity at all three domains, the model underpredicted the total amount of SWE and snow depth at nearly all evaluation sites for the baseline simulations (Figs. 3-5).

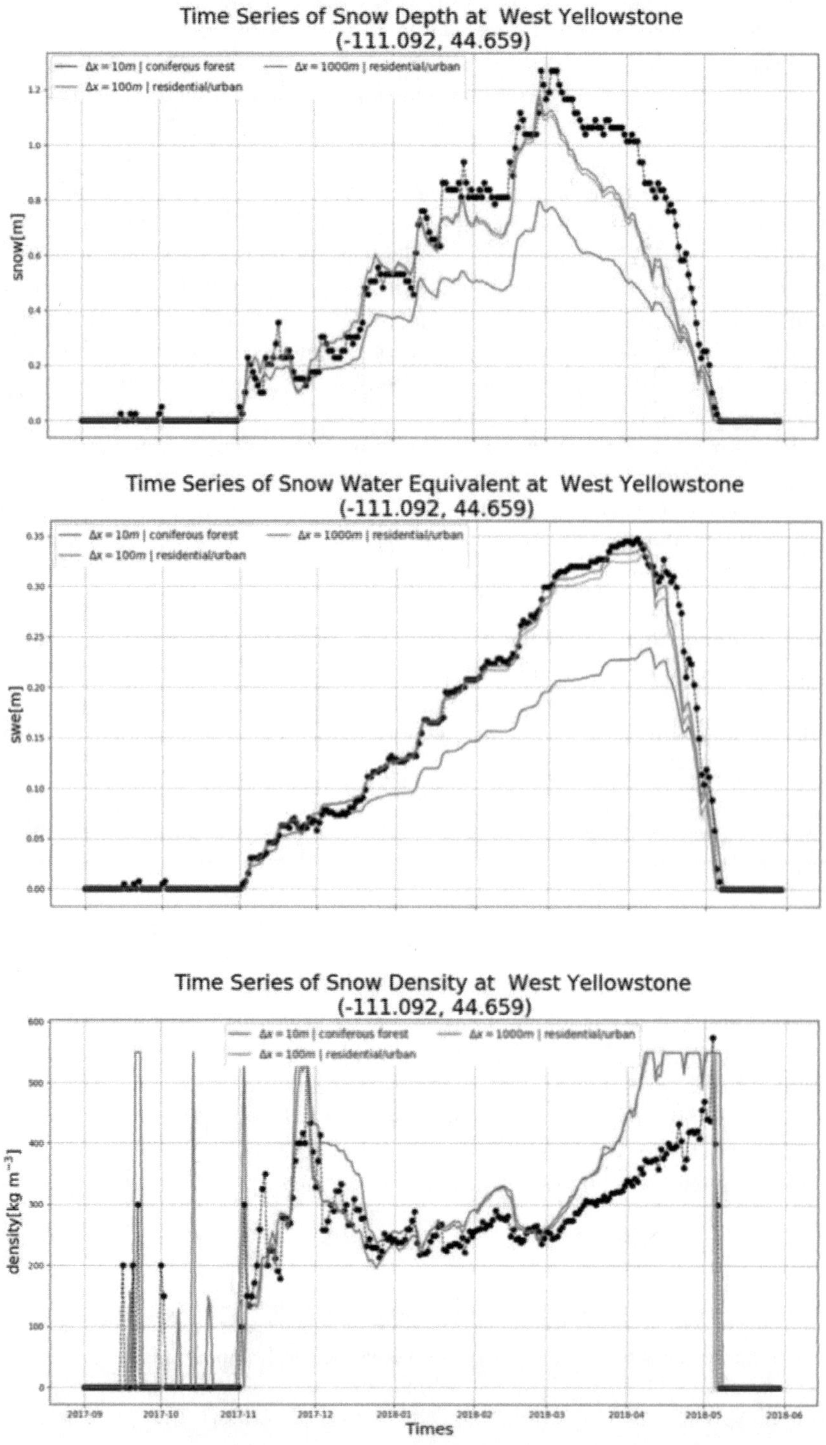

Figure 3. Time series of snow depth, SWE and snow density for the West Yellowstone SNOTEL site.

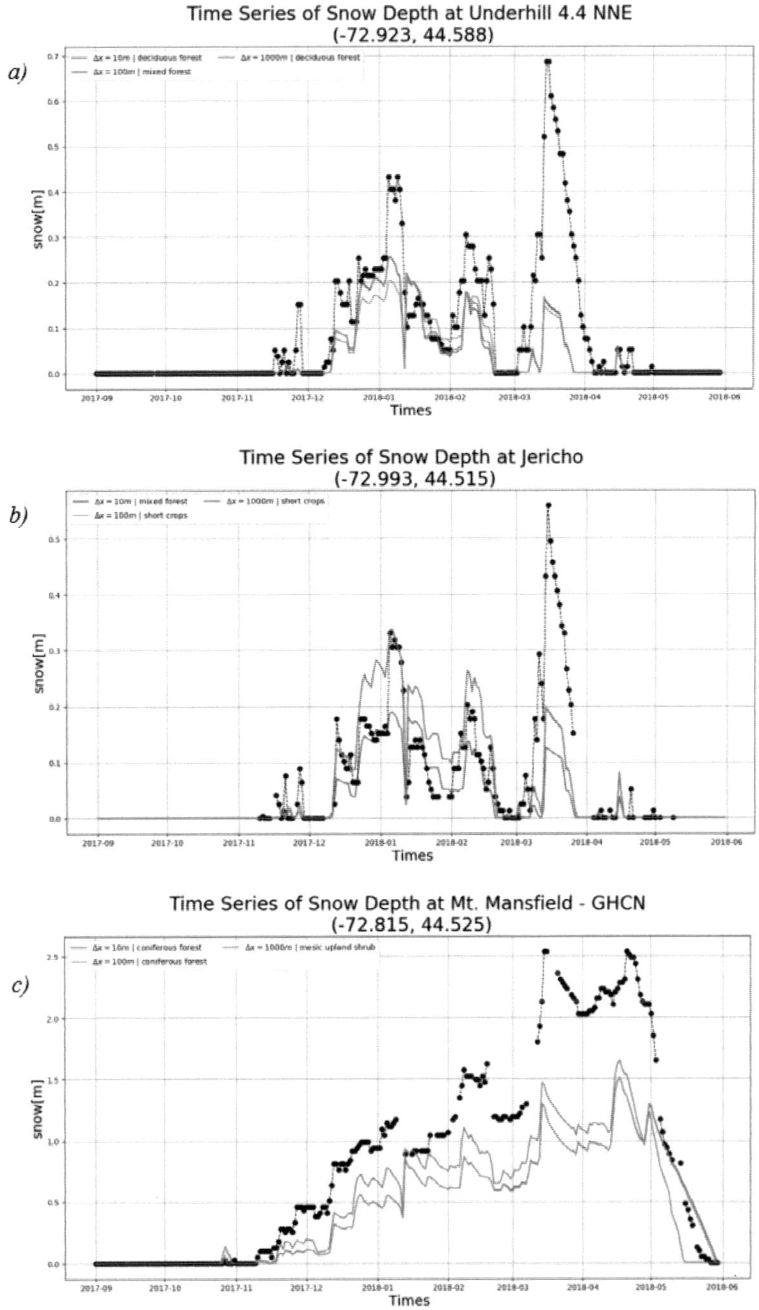

Figure 4. Snow depth, modeled (solid lines) and observed (connected points), at (a) Underhill, VT, (b) Jericho, VT, and (c) Mt. Mansfield for the 2017-2018 snow accumulation and ablation season. The three model resolution runs (10-m, 100-m, and 1000-m) are shown and the resulting land cover classification for each reported in the legend.

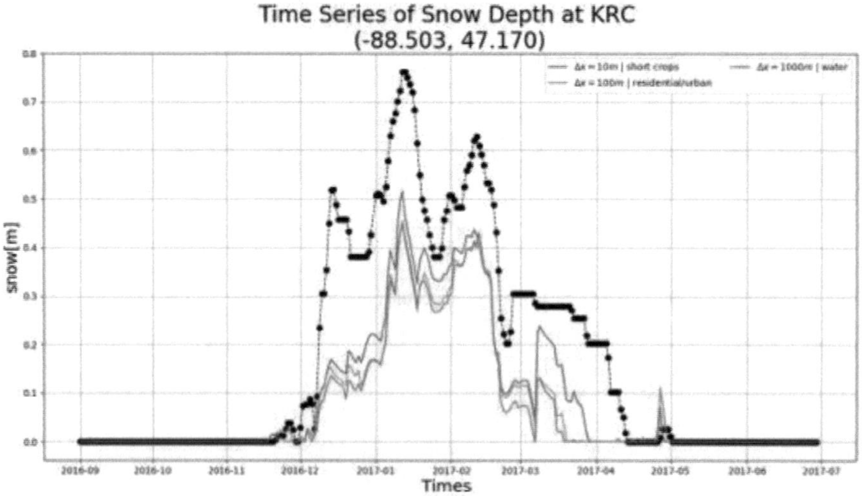

Figure 5. Snow depth, modeled (solid lines) and observed (connected points), at the KRC site for the 2017-2018 snow accumulation and ablation season. The three model resolution runs (10m, 100m, and 1000m) are shown and the resulting land cover classification for each reported in the legend.

All three domains showed qualitatively different snowpack characteristics. For example, the West Yellowstone domain showed a generally consistent accumulation season with a gradual increase in the amount of SWE on the ground, and a slow compaction of the snow. In contrast, the sites within the CEATS domain showed numerous melt events that entirely depleted the snowpack followed by a rapid replenishment of SWE. The exception to this behavior was at the Mt. Mansfield site which is located at an elevation ~4000 feet and is more characteristic of an alpine forest than the typical mid-latitude temperate deciduous forest. At KRC, the snowpack was most similar to that of the low-elevation CEATS sites, with occasional mid-winter melt events that interrupted the steady accumulation of snow.

In general, the baseline SnowModel configuration was able to reproduce the snowpack accumulation and melt in all three domains. In particular, the West Yellowstone simulation performed remarkably well with respect to the timing of maximum snow depth and the timing of melt out at each SNOTEL site (Fig. 3). However, snow depth was largely underestimated throughout the year at each site, with the maximum snow depth underestimated by approximately 0.4 and 0.6 m at each site, respectively. Concerningly, the West Yellowstone SNOTEL site is located in close proximity to the meteorological forcing location and is at the same elevation, thereby ruling out improper terrain downscaling as the explanation for the discrepancy. The model also appears to overestimate density during the late winter and early spring when the snow was undergoing compaction. The overestimate in density combined with the underestimate in SWE amplifies the underestimate in snow depth.

Intriguingly, model performance with respect to SWE and snow depth appears to improve as model resolution is degraded from 100-m to 1-km, contrary to expectations. However, upon examination, we find that the improvement stems from the fact that the resolution decrease causes the land classification to change from forested to unforested (Figs. 3-5). This strongly suggests that resolution has a minimal impact on simulated snow relative to the forest canopy impact. Yet, because SNOTEL observation sites are purposefully sited in clearings, it cannot be concluded that the forest canopy parameterizations are inaccurate from this analysis, since the observation is more representative of a non-forested location, despite its model land classification (Meromy *et al.*, 2013).

Model performance for the CEATS domain largely mirrored that of West Yellowstone. The simulation performed well with respect to snow timing, though in most instances underestimated snow depth. Additionally, resolution degradation had a minimal impact on the results except where it resulted in a change from forested to non-forested land cover type.

Model performance at KRC is similar to performance in the other two domains, but a little more nuanced. While the model systematically underestimates snow depth, there is an apparent greater dependence on resolution that is unaccompanied by a change in canopy coverage.

While increased resolution does not appear to degrade model accuracy substantially, land cover reclassifications notwithstanding, one potential benefit of running SnowModel at fine (<100 m) resolution is its potential to accurately simulate fine-scale spatial heterogeneity of snow depth by including snow redistribution by wind.

We further assess the model's ability to simulate fine-scale spatial heterogeneity by comparing simulated snow depth against manually collected snow depth observations in the West Yellowstone training site. At this site, numerous point snow depth measurements were collected within a small area (<1 km^2). These observations are evaluated against simulated snow depth from the 10-m run. This evaluation largely shows that simulated snow depth variability was substantially less than the observed variability, suggesting that running SnowModel at the 10-m DEM resolution does not adequately reproduce the observed spatial variability observed in reality (Fig. 6). We suspect that this is due to the coarse meteorological forcing and simple relationships governing changes in windspeed, temperature, and precipitation with elevation and land cover. Furthermore, in non-forested areas prone to drifting, small turbulent eddies, not simulated in SnowModel, are largely responsible for determining the high-spatial variability in snow depth (e.g., Liston *et al.*, 2018). However, the high-density observations were collected on flat terrain during the mid-winter, and therefore it cannot be determined from this analysis whether or not the terrain-adjustments to incoming shortwave radiation improve the results at finer resolutions. It seems reasonable to believe the higher resolution will provide greater benefits in regions with complex terrain, where local-scale snow amounts are determined primarily by terrain features than by large-scale meteorology. More work is needed to fully assess the benefits of running SnowModel at higher resolutions.

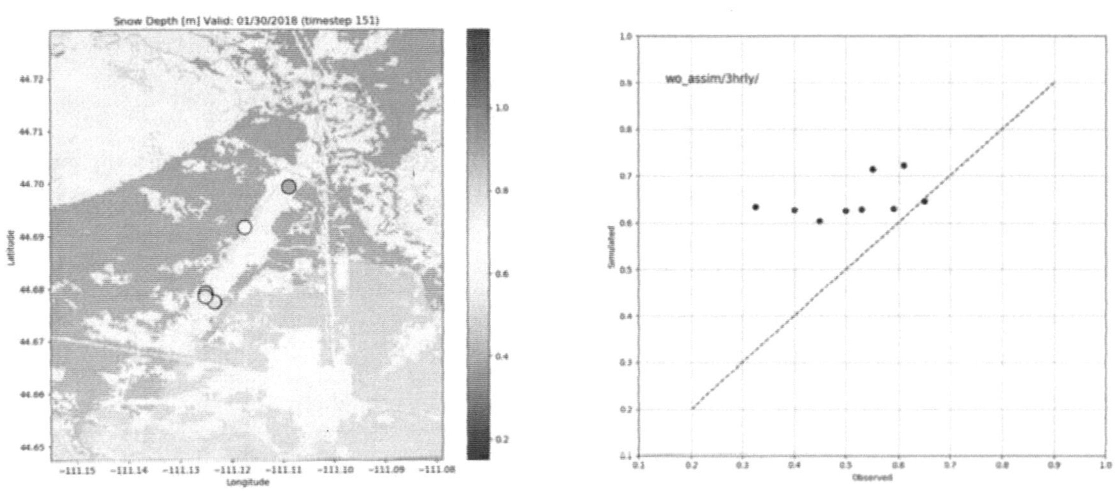

Figure 6. Left: Map of model simulated snow depth with points representing the observations for West Yellowstone domain. Right: Scatter plot of modeled versus observed snow depth from West Yellowstone.

DISCUSSION

The most interesting result from this study was the outsized role that the forest canopy played in governing the snowpack in all three domains. To investigate these impacts further, we examine the canopy parametrizations that affect the snowpack. Specifically, we focus on leaf area index (LAI) which quantifies the fractional leaf area coverage per unit ground area. In SnowModel, LAI is land-category and seasonally dependent following a modified cosine function that computes a LAI value that falls somewhere between winter minimum and summer maximum value. This function is geographically constant, i.e., it has no dependence on latitude or elevation. LAI is used to compute the canopy effects on all driving meteorology at the surface.

In particular, maximum canopy snow holding capacity is computed as:

$$S = 4.4LAI \tag{5}$$

where S is the maximum snow holding capacity. Snow can be removed from the canopy by either unloading or sublimation. Unloaded snow is added to the sub-canopy snowpack. Canopy unloading is determined to be 5 mm day^{-1} °C^{-1} when the air temperature is greater than freezing:

$$U = \begin{cases} \frac{5}{86400}(T_{air} - 273.15)\Delta t, \; T_{air} > 273.15 \\ 0, \; T_{air} < 273.15 \end{cases} \tag{6}$$

where U is the unloaded snow. SnowModel does not include wind unloading. Therefore, snow is only transferred from the canopy to the ground when the air temperature is above freezing. Canopy sublimation is determined using an energy balance approach and is not described in detail here; see Liston and Elder (2006a).

The canopy also has a strong influence on the surface energy budget by modifying the incoming radiative fluxes. Shortwave radiation is attenuated by the canopy by applying a canopy shortwave transmissivity (τ) to Q_s from MicroMet following:

$$\tau = e^{-0.71LAI} \tag{7}$$

and longwave radiation is modified according to:

$$Q_{l,can} = (1 - f_c)LW \downarrow + f_c \sigma T_{air}^4 \tag{8}$$

where, f_c is the longwave forest fraction:

$$f_c = 0.55 + 0.29 \ln(LAI) \tag{9}$$

To illustrate the impact of LAI, the annual cycle of the above variables is plotted for each forest type in SnowModel (Fig. 7). Coniferous forests have the greatest impact on the forcing data that drives SnowModel. Furthermore, unlike deciduous trees, their impacts persist throughout the winter season. This indicates that evergreen forests will have the highest impact on snow interception and radiation.

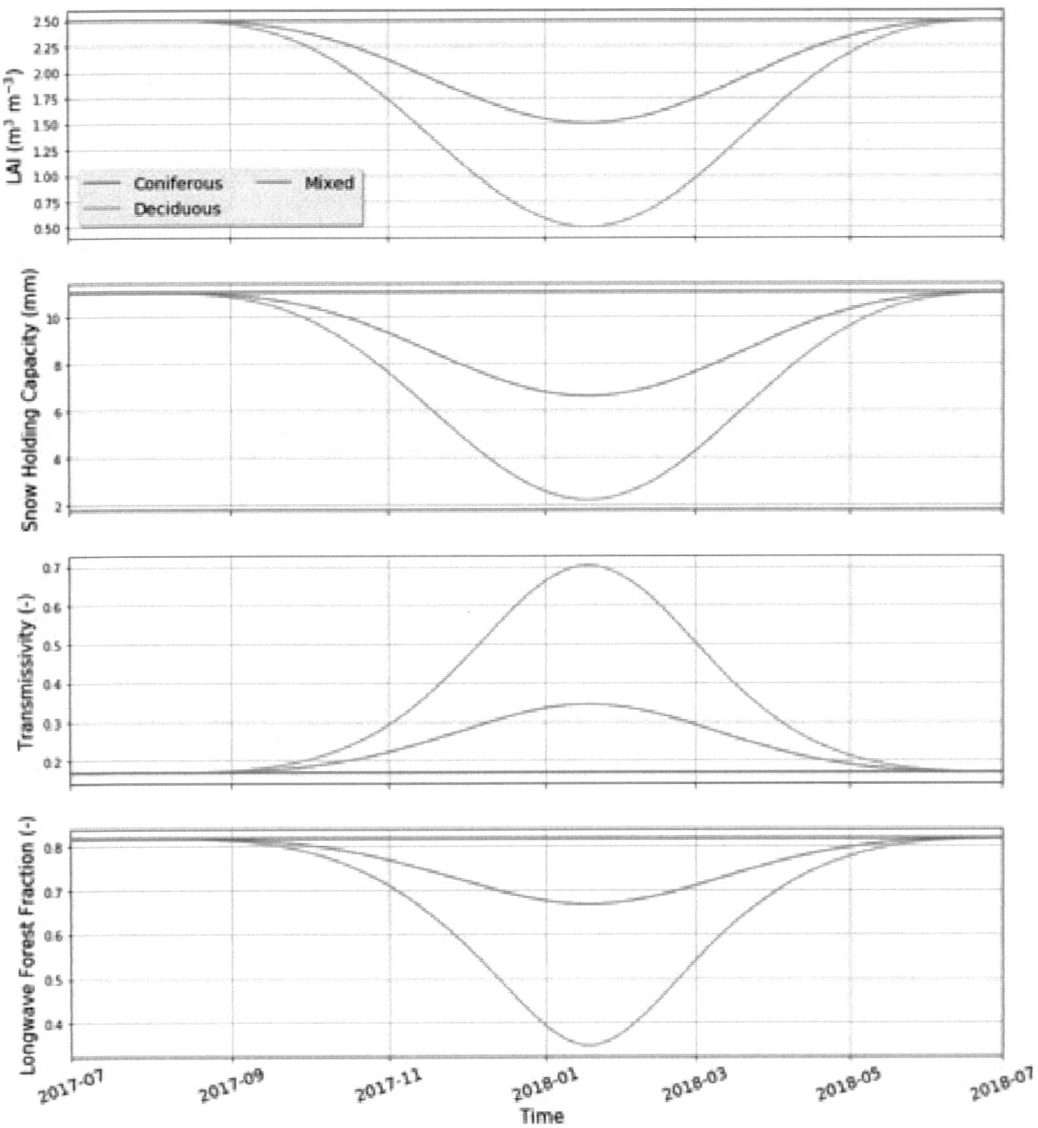

Figure 7. Seasonal cycle of LAI for the three main forest classifications from SnowModel with their impacts on snow holding capacity, shortwave transmissivity, and longwave radiation.

To assess the LAI parameterization, the simulated LAI for each region is computed by weighting each canopy LAI function by its domain fraction. The domain average LAI is then plotted against the MODIS derived LAI (MCD15: Myneni *et al.*, 2016) averaged over the domain (Fig. 8). In general, the SnowModel LAI parameterization does not adequately capture the seasonal cycle of LAI. For instance, the modeled gradual decrease (increase) in LAI during the fall (spring) in the CEATS and KRC domains is at odds with the observed rapid transitions seen in the MODIS data. Furthermore, the assumption that LAI is constant throughout the year for the evergreen forests in West Yellowstone is not representative of the observed seasonal cycle. However, in contrast to SnowModel, MODIS LAI is not a combined LAI and stem area index (SAI) variable. Furthermore, the MODIS LAI product is subject to substantial error for snow covered evergreen trees (Tian *et al.*, 2004). Therefore, caution should be exercised when interpreting differences between the MODIS derived and SnowModel LAI.

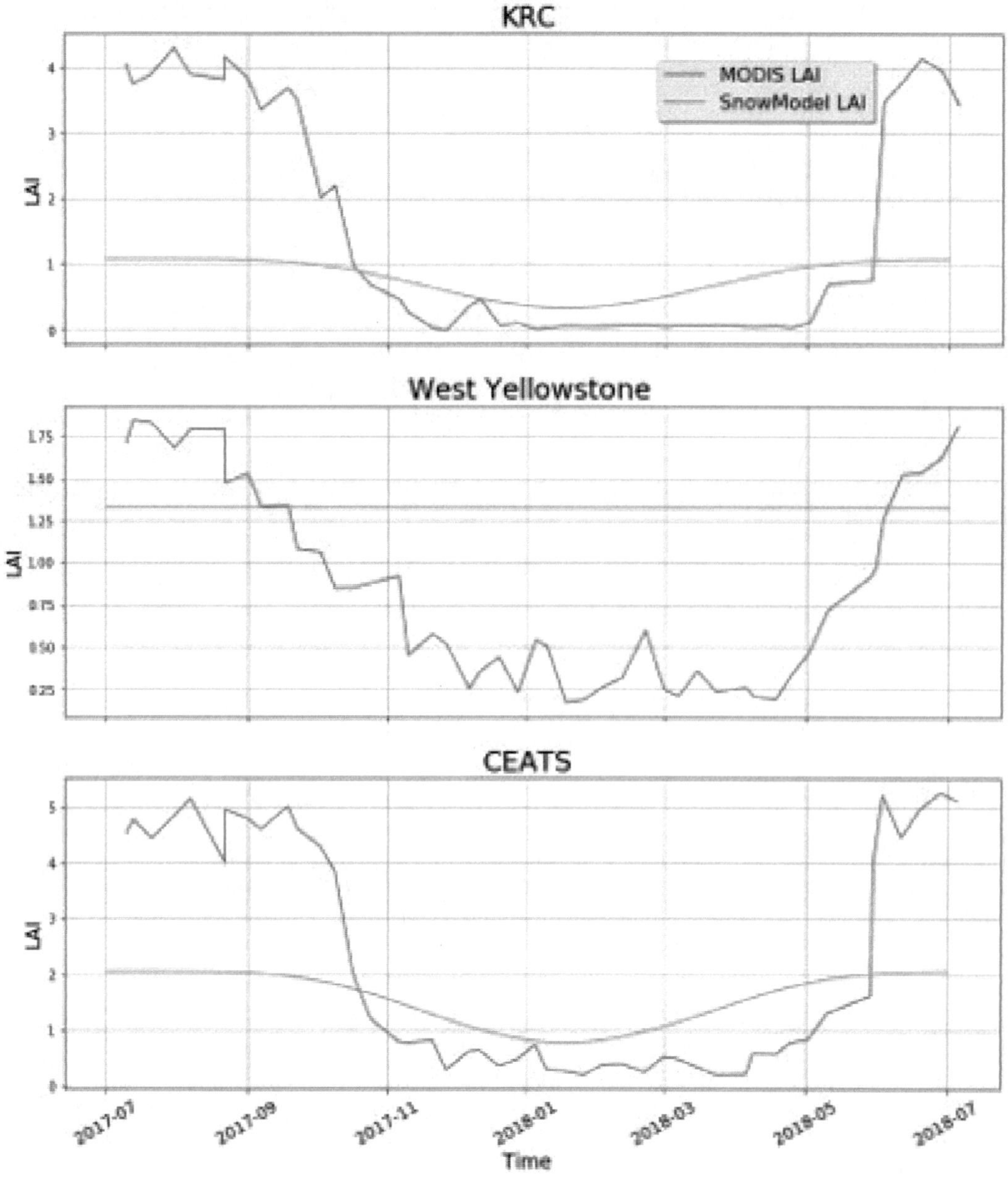

Figure 8. Mean MODIS LAI plotted for each domain against mean LAI from SnowModel. SnowModel LAI is computed by weighting the LAI from all land cover types within each domain.

While we have no direct sub-canopy measurements, we speculate that simulated snow beneath a canopy is subject to large inaccuracies due to its simplified snow unloading parameterization and its overly simplistic representation of the LAI seasonal cycle. We anticipate that these inaccuracies are most likely to lead to a low bias in snow and are likely to be highest for evergreen forests. Combined with the model's general overestimation of snow density, the snow underestimation potentially leads to a substantial overestimate in vehicle mobility. That is, under forest canopies the model produces less, and more dense, snow than in reality.

We note further that the observational comparisons can be disproportionally skewed by the presence of a canopy. Meteorological data is purposefully sited such that it is in a clearing to minimize the impacts of forest on the observations. In particular, this explains the relatively poor performance of SnowModel in the West Yellowstone domain, as gauged by SNOTEL.

Finally, the conclusions of this study are based on a single snow year and the generality of these conclusions is unclear. For example, preliminary evaluation of a SnowModel simulation performed over the CEATS domain during the 2018-2019 season shows much less agreement with available observations, indicating that more work is required to fully understand SnowModel performance in temperate forests.

CONCLUSIONS

In general, SnowModel is able to reproduce the seasonal cycle of snow depth in most instances with reasonable fidelity for numerous different snowpacks, indicating that it has the potential to be a valuable tool for assessing and predicting vehicle mobility in snow covered environments. Limited snow density measurements indicate that the model tends to overestimate snow density in the late-winter and spring. However, density was only evaluated at the West Yellowstone site, so it is unclear as to whether or not this finding is applicable to snowpacks in other regions.

We find that model resolution has a nearly indiscernible direct impact on simulated snow in all three domains between 10-m – 1-km. This is unsurprising in that the meteorological forcing is essentially uniform throughout each domain, and effects of terrain are muted since in all domains, the terrain variability is generally moderate enough such that the model elevation only varies ~100-200 m between the coarse and fine resolutions. Furthermore, because all three domains have substantial tree cover, wind redistribution of snow is limited. However, it is somewhat surprising that neglecting sub-grid snow cover fraction at 1-km resolution had almost no direct impact on the results.

While resolution did not directly affect the results, land cover changes due to down sampling high resolution data had a substantial impact. The effects were particularly significant for grid cells that changed from forested to non-forested or *vice versa* during the down sampling process.

Further investigation into the overwhelming impact of land cover revealed that the seasonal cycle of LAI is likely misrepresented by SnowModel for all major forest types such that canopy effects on snow holding and radiation are likely overestimated throughout the winter. Overestimated canopy effects are likely greatest for evergreen forests with long periods of time of below-freezing snow temperatures. Based on this analysis, caution should be exercised when evaluating modeled snow mobility impacts at grid cells classified as forested. Specifically, we speculate that vehicle mobility is subject to being overestimated for forested land cover types due to a high bias in simulated snow density, and a low bias in simulated snow amount. Combined, these biases lead one to believe that the under-canopy snow is denser and less deep than it likely is in reality.

SnowModel has the ability to "back-correct" an assumed precipitation undercatch observation deficiency using observed snow depth through a form of data assimilation. The results presented here indicate that a high degree of caution needs to be exercised when employing this technique, to ensure that the precipitation correction is being applied correctly. For example, instances when the snow depth used in the assimilation process is located at a grid cell classified as forest canopy, it has the potential to overcorrect the precipitation leading to vastly overestimated snow depth at all proximity grid cells classified as non-forested.

It is possible that the SnowModel canopy parametrization could be improved by updating the LAI seasonality using new relationships available from more recent satellite missions (e.g., MODIS), and by incorporating more sophisticated snow unloading parameterization. These improvements would likely lead to greater confidence in vehicle mobility predictions derived from SnowModel.

REFERENCES

Birkeland KW, Hansen KJ, Brown RL. 1995. The spatial variability of snow resistance on potential avalanche slopes. *Journal of Glaciology*, **41**(137): 183-190.

Blaisdell GL, Richmond PW, Shoop SA, Green CE, Alger RG. 1990. Wheels and Tracks in Snow. Validation Study of the CRREL Shallow Snow Mobility Model (No. CRREL-90-9). Cold Regions Research and Engineering Laboratory, Hanover NH.

Brun E, Rey L. 1987. Field study on snow mechanical properties with special regard to liquid water content. IAHS Publication, 162, pp.183-193.

Iziomon MG, Mayer H, Matzarakis A. 2003. Downward atmospheric longwave irradiance under clear and cloudy skies: Measurement and parameterization. *Journal of Atmospheric and Solar-Terrestrial Physics*, **65**(10): 1107-1116.

Jellinek HHG. 1959. Compressive strength properties of snow. *Journal of Glaciology*, **3**(25): 345-354.

Lever JH, Denton D, Phetteplace GE, Wood SD, Shoop SA. 2006. Mobility of a lightweight tracked robot over deep snow. *Journal of Terramechanics*, **43**(4): 527-551.

Liston GE, Elder K. 2006a. A distributed snow-evolution modeling system (SnowModel). *Journal of Hydrometeorology*, **7**(6): 1259-1276.

Liston GE and Elder K. 2006b. A meteorological distribution system for high-resolution terrestrial modeling (MicroMet). *Journal of Hydrometeorology*, **7**(2): 217-234.

Liston GE, Sturm M. 1998. A snow-transport model for complex terrain. *Journal of Glaciology*, **44**(148): 498-516.

Liston GE, Polashenski C, Rösel A, Itkin P, King J, Merkouriadi I, Haapala J. 2018. A Distributed Snow-Evolution Model for Sea-Ice Applications (SnowModel). *Journal of Geophysical Research: Oceans*, **123**(5): 3786-3810.

Meromy L, Molotch NP, Link TE, Fassnacht SR, Rice R. 2013. Subgrid variability of snow water equivalent at operational snow stations in the western USA. *Hydrological Processes*, **27**(17): 2383-2400.

Myneni R, Knyazikhin Y, Park T. 2016. MOD15A2H MODIS Leaf Area Index/FPAR 8-Day L4 Global 500m SIN Grid V006. NASA EOSDIS Land Processes DAAC.

Pielmeier C, Schneebeli M. 2003. Stratigraphy and changes in hardness of snow measured by hand, ramsonde and snow micro penetrometer: a comparison with planar sections. *Cold Regions Science and Technology*, **37**(3): 393-405.

Peterson TC, Vose RS. 1997. An overview of the Global Historical Climatology Network temperature database. *Bulletin of the American Meteorological Society*, **78**(12): 2837-2850.

Pytka J. 2010. Determination of snow stresses under vehicle loads. *Cold Regions Science and Technology*, **60**(2): 137-145.

Shoop SA, Richmond PW, Lacombe J. 2006. Overview of cold regions mobility modeling at CRREL. *Journal of Terramechanics*, **43**(1): 1-26.

Smith A, Lott N, Vose R. 2011. The integrated surface database: Recent developments and partnerships. *Bulletin of the American Meteorological Society*, **92**(6): 704-708.

Tian Y, Dickinson RE, Zhou L, Zeng X, Dai Y, Myneni RB, Knyazikhin Y, Zhang X, Friedl M, Yu H, Wu W. 2004. Comparison of seasonal and spatial variations of leaf area index and fraction of absorbed photosynthetically active radiation from Moderate Resolution Imaging Spectroradiometer (MODIS) and Common Land Model. *Journal of Geophysical Research: Atmospheres*, **109**(D1).

Walcek CJ. 1994. Cloud cover and its relationship to relative humidity during a springtime midlatitude cyclone. *Monthly Weather Review*, **122**(6): 1021-1035.

Synthetic Comparisons of Snow Observation Constellation Configurations

BARTON A. FORMAN[1], SUJAY KUMAR[2], JONATHAN P. VERVILLE[3], JOSEPH E. GURGANUS[3], LIZHAO WANG[1], JONGMIN PARK[1], AND JAWAIRA AHMAD[1]

ABSTRACT

Obtaining an accurate, global picture of snow mass has proven to be a challenge, in part because no single type of observation or retrieval algorithm works for all types of snow under all conditions. One solution is to combine different observation types – passive microwave, active microwave, passive optical, and active thermal – and to merge those observations with a land surface model in order to synthesize a global snow mass product. Sensors already in orbit, plus sensors planned to be launched in the future, can be merged to explore a multitude of different information mixes, including LiDAR, RADAR, and radiometry. Within this mix of different sensors is a complex tradespace involving swath width, repeat interval, orbit inclination, footprint size, footprint spacing, observation accuracy, and error characteristics.

The study presented here utilizes NASA's Land Information System (LIS) in conjunction with the Tradespace Analysis Tool for Constellations (TAT-C) to explore potential combinations of existing and future sensors. For a given orbital configuration and mix of sensors, these simulations help quantify how much of the seasonal snow world can be observed, how often, with what footprint size and spacing and with what swath width. Such information will be highly valuable for informing discussions on future snow mission concepts. It will also highlight where modeling efforts can provide the greatest impact and perhaps indicate the parameters needing the greatest improvements in accuracy or precision. The results of the simulations will help make progress toward accurate global snow products.

[1] Department of Civil and Environmental Engineering, University of Maryland, College Park, MD, USA
[2] Hydrological Science Laboratory, NASA Goddard Space Flight Center, Greenbelt, MD, USA
[3] Software Engineering Division, NASA Goddard Space Flight Center, Greenbelt, MD, USA

Merging Regional Climate Models and Remote Sensing Datasets to Estimate Mountain Snow Water Equivalent: Proof-of-Concept in the Tuolumne Watershed

MICHAEL T. DURAND[1,2], MELISSA L. WRZESIEN[3], JESSICA LUNDQUIST[4], LAURA HINKELMAN[5], KARL RITTGER[6], JEFF DOZIER[7], TAMLIN M. PAVELSKY[3], SARAH B. KAPNICK[8], AND KRISTEN RASMUSSEN[9]

ABSTRACT

Large-scale high-resolution estimation of snow water equivalent (SWE) in mountainous areas is challenging. Two approaches currently deployable at continental scale are SWE reconstruction and regional climate model (RCM) simulation. Here, we present a method that produces a simultaneous estimate of daily mass and energy balances at 500 m resolution, including SWE timeseries, informed by RCMs and constrained by observations in a way similar to SWE reconstruction. We formulate this as a constrained optimization problem; we seek to minimize the difference between our estimates and observed MODIS snow-covered fraction (SCF) and CERES irradiance, as well as RCM SWE from 3-km Weather Research and Forecasting (WRF) model simulations, subject to mass and energy balances constraints. This problem is readily solved using off-the shelf software. We compute Tuolumne watershed SWE (where it flows into the Hetch Hetchy reservoir: 775 km^2 or 3,612 MODIS pixels) in the Sierra Nevada, USA for water year 2009, a year with average snow accumulation. We validate against snow pillows and snow course data. We find that the SCF and irradiance observations constrain the WRF estimates significantly, with final RMSE of 66 mm and 98 mm at two snow pillows within the watershed, about 15% of peak SWE. Across the watershed, the total SWE volume estimated by our algorithm (0.34 km^3) compared well to high-resolution (90 m) SWE reconstruction (0.38 km^3), while WRF alone was too high (0.45 km^3). Our method represents a compromise, leveraging the beneficial qualities of both RCMs and reconstruction, and producing a simultaneous estimate of mass and energy fluxes and storages applicable to mountain regions.

[1] School of Earth Sciences, The Ohio State University, Columbus, OH, USA
[2] Byrd Polar and Climate Research Center, The Ohio State University, Columbus, OH, USA
[3] Department of Geological Sciences, University of North Carolina, Chapel Hill, NC, USA
[4] Department of Civil and Environmental Engineering, University of Washington, Seattle, WA, USA
[5] Joint Institute for the Study of the Atmosphere and Ocean, University of Washington, Seattle, WA, USA
[6] Institute of Arctic and Alpine Research, University of Colorado, Boulder, CO, USA
[7] Bren School of Environmental Science and Management, University of California, Santa Barbara, CA, USA
[8] Geophysical Fluid Dynamics Laboratory, National Oceanic and Atmospheric Administration, Princeton, NJ, USA
[9] Department of Atmospheric Science, Colorado State University, Fort Collins, CO, USA

Effects of Harvesting and Vegetation Change on Snow Accumulation and Melt in Boreal Forest

MAXIME BEAUDOIN-GALAISE[1] AND SYLVAIN JUTRAS

ABSTRACT

For the Boreal Forest in eastern Canada, previous studies related to the impact of logging on hydrological processes have focused on analyzing short-term changes using empirical relationships and paired-basin approach. In order to contribute to the improvement of knowledge in forest hydrology, the main objective of this research project is to analyze the effects of harvesting and vegetation change on boreal forest water balance and runoff. Given the importance of snow processes on the annual water balance, the first objective of this thesis is to evaluate the effect of logging and regeneration on snow accumulation and melt rate using a modeling approach. The hydrological model chosen for this study is the Cold Region Hydrological Model (CRHM). Based on the long history of logging and hydro-meteorological data from the Bassin Expérimental du Ruisseau des Eaux- Volées (BEREV) at Forêt Montmorency, this study will be one of the first implementation of CRHM in the province of Quebec. With manual snow measurements since 1965, the CRHM model will be parameterized and validated on a multitude of meteorological events. Based on previous studies of the relationship between forest cover and snow processes at Forêt Montmorency and the physically-based algorithms in CRHM, the model is expected to perform well. With good confidence in the model's ability to simulate snow processes, the parameterization of the model will be extended to the building of a complete hydrological model to assess the long-term effect of logging and vegetation changes on streamflow.

[1] Department of Wood and Forest Sciences, Université Laval, Québec City, QC, Canada

Future Changes in Mean and Extreme Daily Snowfall over the United States

RACHEL R. MCCRARY[1], JENNIFER M. JACOBS[2], AND LINDA O. MEARNS[1]

ABSTRACT

The characteristics of snowfall and the subsequent accumulation of winter snowpacks across the continental United States (US) will change in a warmer climate. These changes will have many societally relevant implications by influencing water supply and flood management, recreation, ecosystem health and diversity, and the occurrence of hazardous and damaging snowstorms. Snow is a normal part of the climate for most regions of the US, so these changes will potentially affect a significant fraction of the US population. It is therefore critical that US decision makers are armed with integrated projections of long- term climate changes affecting snowfall for the entire US when preparing for future climate conditions.

This study explores how anthropogenic climate change is projected to influence snowfall characteristics (timing, amount, and intensity) over the continental United States (CONUS) using a suite of dynamically downscaled, high-resolution (12km, 25km, and 50km) climate change simulations from the North America Coordinated Regional Climate Downscaling Experiment (NA-CORDEX) and the Framework for Assessing Climate's Energy-water-land nexus using Targeted Simulations (FACETS) ensembles. Analysis includes an evaluation of the historical climate simulations as well as an estimation of future changes in mean and extreme snowfall for all of CONUS.

[1] National Center for Atmospheric Research, Boulder, CO, USA
[2] University of New Hampshire, Durham, NH, USA

Measurement of Tundra Arctic Snow Microstructure and Improved Microwave Radiometry Modelling

CÉLINE VARGEL[1,2,3], ALAIN ROYER[1,2], VINCENT SASSEVILLE[1,2], OLIVIER SAINT-JEAN RONDEAU[1], GHISLAIN PICARD[3], ALEXANDRE LANGLOIS[1,2], AND ALEXANDRE ROY[2,4]

ABSTRACT

Arctic snow has the peculiarity of being very dense near the surface, due to frequent blowing-snow events and sustained cold temperatures, while the bottom of the snowpack is typically less dense, with thick depth hoar layers formed through temperature gradient metamorphism. This leads to a combination of high thermal conductivity contrast, where the upper layers have higher conductivity than the bottom layers which contains more air. The result of this combination can significantly alter snowpack-insulating properties, as these layers develop through the winter. Currently, no land surface model is able to accurately simulate such a specific density stratigraphy. Snow models generally produce the reverse density stratigraphy, with low density at the surface and high snow density at the bottom since they are mostly based on compaction and ignore the water vapor fluxes from the ground as well as the vegetation interactions, both leading to depth hoar formation.

We show this issue can be resolved using passive microwave radiometry. From simulations based on the Snow Microwave Radiative Transfer (RT) model SMRT, we discuss how the arctic snow brightness temperature (TB) can depart from temperate and subarctic snow due to the different microstructures and stratigraphy of snowpacks. Based on new ground-based radiometric and snow microstructure measurements over Canadian Arctic tundra sites (Umiujaq, Nunavik, Trail Valley Creek, Yukon, and Cambridge Bay, Nunavut, Canada), we parameterize the snowpack microstructure profile using snow surface specific area, density and micropenetrometer-derived correlation length variations. We then discuss the best modelling and configuration approach among the RT models proposed in SMRT (DMRT-SHS, IBA-SHS and IBA-Exp). The presence of wind slab and ice crusts within the snowpack, which are now often observed over Arctic areas, also generates significant variation in TB.

These results suggest that improved tundra arctic snow modelling could lead to better estimates of snow insulating effects on the ground and thus reduces errors in estimates on winter soil temperature and permafrost evolution (Arctic snow feedback effect).

[1] Centre de Recherches et d'Applications en Télédétection (CARTEL), Université de Sherbrooke, Sherbrooke, QC, Canada
[2] Centre d'Etudes Nordiques, Québec City, QC, Canada
[3] Institut des Géosciences de 'Environnement, Université Grenoble Alpes – CNRS, Grenoble, France
[4] Université du Québec à Trois-Rivières, Trois-Rivières, QC, Canada

Dust Associated Microorganisms and Impacts on Snow Melt and Snow Structure

ALISON K. THURSTON[1], LAUREN B. FARNSWORTH[1], JOHN M. FEGYVERESI[1], ROSS LIEBLAPPEN[1], STACEY L. JARVIS[1], SHELBY A. ROSTEN[1], ROBYN A. BARBATO[1], AND ZOE R. COURVILLE[1]

ABSTRACT

Dust is transported onto snow covered regions either via wind redistribution (dry deposition) or from the atmosphere during a snowfall event (wet deposition). Dust particles carry microbial and chemical signatures from the dust source region to the deposition region. Microorganisms become incorporated into, and can greatly alter, snowpack physical properties including snow structure, pore structure and resulting radiative and mechanical properties. These processes affect the surrounding hydrology on a macro-scale. In this interdisciplinary study, we examined microbial-associated dust-dependent melt effects on snow melt and snow strength predictions. Our ultimate goal was to determine if we could find unique microbial communities according to a dust event to eventually attempt to characterize microbial signatures.

[1] U.S. Army Engineer Research and Development Center (ERDC) Cold Regions Research and Engineering Laboratory (CRREL), Hanover, NH, USA

Dust on Snow Impacts to Alpine Areas

LAUREN FARNSWORTH[1], ROBYN BARBATO[1], ALISON THURSTON[1], ZOE COURVILLE[1], AND ROSS LIEBLAPPEN[1,2]

ABSTRACT

Dust is transported onto snow covered regions either via wind redistribution or from the atmosphere during a snowfall event. Dust particles carry microbial and chemical signatures from the dust source region to the deposition region. Microorganisms become incorporated into, and can greatly alter, snowpack physical properties including snow structure, pore structure, and resulting radiative and mechanical properties. These processes affect the surrounding hydrology on a macro-scale. In this interdisciplinary study, we examine microbial deposition on alpine snow through dust transport and the effects this deposition has on the snow matrix with the goal of further understanding microbial-associated dust-dependent melt effects on snow melt and snow strength predictions. Our research objectives are to examine the provenance of the dust and associated microorganisms found in the snowpack during the spring of 2017 during a period of widespread dust deposition events. We used molecular techniques to assess the microorganisms present in the samples and found that location is a driving factor of the snow microbial community, and that specific dust deposition events, originating from slightly different locations, can result in different microbial presence in the deposited snowpack. Microstructural analysis of the dust within the snow matrix suggests that for the case of the merged snow layer from late April that dust resides at the snow/pore interface so that it is open to air for respiration. We found that dust characteristics varied with site and that dust was generally located on the exterior of snow grains.

[1] U.S. Army Corps of Engineers Cold Regions Research and Engineering Laboratory, Hanover, NH, USA
[2] Vermont Technical College, Randolph Center, VT, USA

Observation of the Microstructural Evolution of Polar Firn under Compression in a MicroCT

YUAN LI[1] AND IAN BAKER[1]

ABSTRACT

The aim of this work is to understand how compression impacts the densification of polar firn. We applied a compressive load using a Material Testing Stage in a microcomputed X-ray tomograph (microCT) located in a cold room at -10 °C to samples taken at ~ 10 m intervals along the length of 80 m firn core extracted at Summit, Greenland in July 2017. Each sample was intermittently compressed in increasing strain increments at a strain rate of ~ 8×10^{-5} s^{-1}. Several features are noteworthy. First, densification along the whole length of the firn core is accompanied by decreases of the specific surface area, and both total and open porosities with increasing strain. Second, during densification an increase of structure thickness, a measure of the particle or grain aggregate size, occurs with increasing strain. Third, the decreases of the structure model index and surface convexity index that occur with increasing strain imply that the consolidation of particles occurs with increasing strain. Finally, as might be expected, the effect of a compressive load decreases with increasing depth.

This work was sponsored by the National Science Foundation Arctic Natural Science grant 1743106.

[1] Thayer School of Engineering, Dartmouth College, Hanover, NH, USA

Trend and Design of Annual Maximum Snowmelt Events over the Conterminous United States (CONUS)

EUNSANG CHO[1,2] AND JENNIFER M. JACOBS[1,2]

ABSTRACT

Snow impacts on human activity across the U.S. In the north-central and north-eastern U.S., snow meltwater is a dominant driver of severe spring flooding. Recent snowmelt floods in 1997, 2009, 2011, and 2019 resulted in large societal and economic impacts on communities in the north-central and north-eastern U.S. Due to the lack of reliable long-term gridded SWE, the current engineering design precipitation U.S. maps are based solely on rainfall (point interpolation or gridded) data (e.g. NOAA Atlas 14). In this study, we quantify the trends in annual maximum snow water equivalent (SWE) and snowmelt events over the CONUS using observation-based long term 4-km gridded SWE data developed by University of Arizona (UA SWE) from 1981 to 2017. In most mountain regions in western U.S., annual maximum SWE decreased significantly ($p < 0.05$) while there are no significant changes but general increases at a non-significant level, in the north-central and eastern U.S. However, annual maximum 7-day snowmelt increased significantly in north-central U.S. (North Dakota and Minnesota) and parts of Michigan and Maine (7 - 10 mm per decade). Based on the accuracies of the UA SWE products without systematic biases (Cho *et al.*, 2019), engineering design snowmelt maps (25- and 100-year) were firstly generated over the CONUS. We expect that the design maps will help infrastructure design regarding flood risk management in snowmelt dominant regions.

[1] Department of Civil and Environmental Engineering, University of New Hampshire, Durham, NH, USA
[2] Earth Systems Research Center, Institute for the Study of Earth, Oceans, and Space (EOS), University of New Hampshire, Durham, NH, USA

Snowmelt Processes on Antarctic Sea Ice Observed by Radar Scatterometers

STEFANIE ARNDT[1] AND CHRISTIAN HAAS[1]

ABSTRACT

Snowmelt processes on sea ice are the key drivers determining the seasonal sea-ice energy and mass budgets. Around Antarctica, snowmelt on perennial ice is weak and very different than in the Arctic, with most snow surviving the summer. Here, we compile time series of snowmelt onset dates on perennial Antarctic sea ice from 1992 to 2014 using active microwave observations from European Remote Sensing Satellite (ERS-1/2), Quick Scatterometer (QSCAT) and Advanced Scatterometer (ASCAT) radar scatterometers. Describing snow melt processes, we define two transition stages: a weak backscatter rise indicating the initial warming and metamorphosis of the snowpack (pre-melt), followed by a rapid rise indicating the onset of thaw-freeze cycles (snowmelt). Results show large interannual variability with average pre-melt and snowmelt onset dates of 29 November and 10 December, respectively, without any significant trends over the study period. Related to different signal frequencies, we show that QSCAT Ku-band (13.4 GHz signal frequency) derived pre-melt and snowmelt onset dates are earlier by 25 and 11 days, respectively, than ERS and ASCAT C-band (5.6 GHz) derived dates. This offset has been considered when constructing the time series. Combining the observed successive timing of melt events retrieved from radar scatterometers with melt onset dates retrieved from 37 GHz passive microwave radiometers, allows us to develop a conceptual model which illustrates how the evolution of seasonal snow temperature profiles affects different microwave bands with different penetration depths. These suggest that future multi-frequency microwave satellite missions could be used to resolve melt processes throughout the vertical snow column.

[1] Alfred-Wegener-Institut Helmholtz-Zentrum für Polar-und Meeresforschung, Bremerhaven, Germany

Evaluation of Satellite-Derived Estimates of Lake Ice Cover Timing on Svalbard using *in situ* Data

SAMUEL E. TUTTLE[1], STEVEN ROOF[2], JIN CAO[1], ALAN WERNER[1], GRANT GUNN[3], AND ERIN BUNTING[3]

ABSTRACT

Arctic lakes are sensitive indicators of climate change. In recent decades satellites have greatly increased the capability of monitoring lake ice timing, especially in remote areas. However, satellite observations of lake ice in the Arctic are not often ground-truthed with *in situ* measurements and direct observations, due to the remoteness of much of the region. In this study of Lake Linné, one of the largest lakes on Svalbard, Norway in the North Atlantic region, we use continuously monitored lake water temperature and automated photographs from ground-based cameras to evaluate the ability of satellite platforms to capture lake ice timing and duration. Visible and near infrared surface reflectance data from the Moderate Resolution Imaging Spectroradiometer (MODIS) were used to observe the seasonal change in reflectance of Lake Linné from fall 2003 - spring 2018, and to determine summer ice-off (also called break-up end (BUE)). Microwave backscatter data from Sentinel-1 were similarly used to determine BUE and fall freeze-up (also called freeze-up start (FUS)) from fall 2014 - spring 2018. These estimates were directly compared to twice-daily photographs of the lake, as well as inferred ice cover from lake water temperatures. The analysis indicates that satellite-based ice timing estimates for Lake Linné compare favorably with *in situ* data during the study period. Additionally, the data show that lake ice duration has decreased significantly from 2003 - 2018 in this part of Svalbard, with little change in summer break-up but a trend toward much later fall freeze-up.

[1] Mount Holyoke College, South Hadley, MA, USA
[2] Hampshire College, Amherst, MA, USA
[3] Michigan State University, East Lansing, MI, USA

Witchcraft, Wizardry, and Water: The Intersection of Physics, Electrical Engineering, and Snow Monitoring

PAUL W. NUGENT[1], COOPER P. MCCANN[1], AND AUSTIN W. BEARD[1]

ABSTRACT

NWB Sensors, Inc. has been developing prototype technology for measuring snow water equivalent (SWE), snow liquid water content (LWC), snow depth, and snow density by observing the changes in GNSS signals that have transmitted through the snowpack. Snomonstor™, is a fully electronic fluidless snowpack measurement technology targeting the replacement of antifreeze filled snow-pillows in snowpack measurement. By creating a smaller, accurate, easy-to-maintain, and more economical snow measurement system, a larger number of snow measurement stations to be installed in locations where no snowpack data is currently available. This will allow water-hydrologists to have a higher spatial sampling of snow measurement permitting improved watershed forecasts and water management in order to help meet human water needs.

NWB Sensors built prototype snow sensors and over the past three winters deployed it at two USDA operated Snow Telemetry (SNOTEL) sites, and last winter at three Montana Climate Office operated Montana Mesonet sites. Data collected has been used to develop algorithms that derive snow parameters. Agreement with an adjacent snow pillow was within ±7 mm of SWE during both accumulation and ablation (well below the error of the snow pillow itself). Derived LWC data showed daily melt-freeze cycles as expected.

LWC data has compared reasonably to published studies. However, published literature values for mapping between snow density and LWC and complex index of refraction do not necessarily match with the way in which GPS signals propagate through the snow and semi-empirical models are necessary to obtain good agreement with preexisting methods.

[1] NWB Sensors, Inc., Bozeman, MT, USA

Taku Glacier, Alaska in 2018 Highest Snowline in 70+ Years

MAURI PELTO[1]

ABSTRACT

The Juneau Icefield Research Program (JIRP) has been examining the glaciers of the Juneau Icefield since 1946. The height of the transient snowline (TSL) at the end of the summer represents the annual equilibrium line altitude (ELA) for the glacier, where ablation equals accumulation. Until the NASA Landsat program began, field measurements and aerial observations were the only means to observe the ELA. On Taku Glacier the ELA has been observed annually from 1946-2018. The mean ELA has risen 85 m from the 1946–1985 period to the 1986–2018 period. Mean annual mass balance from 1946-1985 and 1986-2018, with 2018 values being preliminary, were +0.40 m a^{-1} and −0.18 m a^{-1}, respectively, indicative of the snow line rise resulting in cessation of the long-term thickening of the glacier.

In 2018, TSL on July 5 was 900 m, on July 21 was 975 m, on July 30 was 1075 m, on Sept. 16 was 1400 m, and on October 1 was 1425 m. This is the first time since 1946 that the snowline has reached or exceeded 1400 m on Taku Glacier. The 500 m rise from July 5 to Sept 16th occurred in ~73 days. With a balance gradient of ~3.3 mm m^{-1} this represents ablation of 1.65 m w.e. snow. On July 22, a snowpit was completed at 1405 m with 0.93 m w.e, that had lost all snowcover by Sept. 16. This is one of seven snow pits completed in July providing field data to verify ablation rate.

[1] Nichols College, Dudley, MA, USA

Characterization of Near Subsurface Conditions at McMurdo Station, Antarctica

ROSA T. AFFLECK[1], SETH CAMPBELL[2], SAMANTHA SINCLAIR[1], AND KEVIN BJELLA[1]

ABSTRACT

The National Science Foundation has been recently approved to move forward with the major infrastructure rebuilding at McMurdo Station (MCM). Efforts to rebuild MCM require knowledge of geology, ground conditions, and geotechnical information of the ice-cemented layer. Therefore, 200 and 400 MHz ground-penetrating radar (GPR) surveys were collected in McMurdo during January, October, and November of 2015 to detect the active layer, permafrost or massive ice, fill thickness, solid bedrock depth, and buried utilities or construction and waste debris. Soil pits were excavated to collect soil, ice, and rock samples for gradation, density, and moisture content tests. Frozen cores were collected in various locations using a chilled air drilling system. Information extracted from the soil pits and the cores were used to corroborate the GPR profiles. The studies revealed distinct features, including ice-bonded fractured basaltic boulders, rocks, and gravelly sand; massive ice; constructed (friable) and contaminated fill layers. A considerable amount of near-surface excess ice found was likely due to the anthropogenic origins from runoff draining and refreezing. If the foundation is placed in contact with the ground allowing heat transfer to take place, removal of contaminated materials and ground ice is recommended so that a new structural base layer is constructed with suitable (processed on-site) fill materials at optimum moisture content and compaction. Recommendations from these studies would allow the benefits in lower construction material needs, increased energy efficiency, minimized drainage issues, and snow/ice accumulation prevention for the new infrastructure at MCM.

[1] U.S. Army Engineer Research and Development Center (ERDC), Cold Regions Research and Engineering Laboratory (CRREL), Hanover, NH, USA
[2] School of Earth & Climate Sciences and Climate Change Institute, University of Maine, Orono, ME, USA

Town Hall / Panel

Session Moderators: Ed Kim and Michael Durand

Ingredients for a Future Snow Satellite Mission

EDWARD KIM[1], MICHAEL DURAND[2], JARED ENTIN[3], BARTON FORMAN[4], DOROTHY HALL[1,4], PAUL HOUSER[5], DO-HYUK KANG[1,4], SUJAY KUMAR[1], JESSICA LUNDQUIST[6], LEUNG TSANG[7], AND CARRIE VUYOVICH[1]

ABSTRACT

Recent work by the snow remote sensing community through SnowEx field campaigns and related efforts supported by NASA's Terrestrial Hydrology Program has begun to refine the ingredients for a future global snow satellite mission. While the outlines of the ingredient list are well-known based on other Earth sensing missions, the detailed contents are still being developed. Indeed, the details of the future SnowEx campaigns are still being developed. Yet this is an opportune moment to take stock of what we think we know, what we still need to determine, and how or why we think so, because the plans have not yet been finalized.

While the details of a snow satellite are still not fully clear, requirements are becoming less fuzzy. The lists of possible sensing techniques and algorithm strengths and limitations are becoming more refined for each type of snow. And since no single technique works for all types of snow, lists of existing and planned spaceborne sensors are being collected with an eye toward considering a constellation approach. The science and application utility of these sensors are being examined through formal observation system simulation experiments. Candidate snowpack, land surface, and radiative transfer models are being explored.

We will provide examples from the list of ingredients, identify key remaining questions or gaps, and suggest ways to address them via field measurements and/or modeling studies. By doing so, we hope to stimulate broad discussion in order to refine the list of ingredients and move closer to a snow mission.

[1] NASA Goddard Space Flight Center, Greenbelt, MD, USA
[2] The Ohio State University, Columbus, OH, USA
[3] NASA Headquarters, Washington, DC, USA
[4] University of Maryland, College Park, MD, USA
[5] George Mason University, Fairfax, VA, USA
[6] University of Washington, Seattle, WA, USA
[7] University of Michigan, Ann Arbor, MI, USA

Session #4: In Situ Snow Observations

Development of an Open-Ended Coaxial Probe (OECP) for Snow Liquid Water Content Measurement

ALEX MAVROVIC[1,2,3], JEAN-BENOIT MADORE[1,2], ALEXANDRE LANGLOIS[1,2], ALAIN ROYER[1,2], AND ALEXANDRE ROY[2,3]

ABSTRACT

Liquid water content (LWC) in snow is an important metric for the evaluation of various snowpack physical processes. The percolation of water from rain and snowmelt can lead to instability through the additional weight of wet snow and the creation of ice layers/crusts. The understanding of those instabilities through percolation schemes as well as their detection is critical in avalanche risk assessment. In addition, rain on snow events occurrence is increasing in Arctic regions and snow LWC greatly affects the microwave signal used to retrieve snow and ground information over vast and remote regions. Precise LWC measurements are required to take into account those particular snow conditions in radiative transfer models in order to improve passive microwave satellite products.

Our team at the University of Sherbrooke has developed a probe to measure the LWC of snow. The approach is based on the strong relationship between snow microwave permittivity (at 1.4 GHz) and LWC knowing that water permittivity is much higher than air and ice permittivity. The performance of the probe is similar to the performance of other instruments currently available but is slightly more accurate. This new instrument is able to quantify the water content of thin layers of percolation accumulation due to its smaller probed volume, which other instruments cannot. The characterization of those thin percolation layers is critical for the validation of percolation models and avalanche risk assessment.

[1] Centre d'Applications et de Recherches en Télédétection, Université de Sherbrooke, Sherbrooke, QC, Canada
[2] Centre d'Études Nordiques, Université Laval, Quéebec City, QC, Canada
[3] Université du Québec à Trois-Rivières, Trois-Rivières, QC, Canada

A (Simple) Probabilistic Approach for Solid Precipitation Undercatch Adjustment

AMANDINE PIERRE[1], FRANCOIS ANCTIL[2], AND SYLVAIN JUTRAS[1]

ABSTRACT

The undercatch of solid precipitation is bound to the uncertainty of the gauge's measurement. This is recognized as a complex phenomenon, which is often adjusted using deterministic transfer functions depending mainly on wind speed and air temperature. The deterministic approach cannot consider or quantify the uncertainty inherent to each meteorological event. This presentation proposes an original mix between calibration of deterministic equations and probabilistic application that enables: i) to quantify the error and variance related to the precipitation gauge measurement ; ii) to validate that error and variance both increase with the wind speed when recording solid precipitation events ; iii) to rank the wind shield types efficiency for different wind speed levels. Data comes from the Neige meteorological station hosted in Montmorency Forest, from 2014 to 2018.

The data, collected automatically and manually by several gauges, are aggregated by type of windshields (Bush, single-Alter, double-Alter, Tretiakov, Nipher, and unshielded devices). Results showed first that the unshielded dataset seemed not large enough to be considered representative of the solid precipitation undercatch phenomenon. Second, 70% (standard deviation ± 3%) of the adjustments made using the probabilistic method corresponded to the reference.

Finally, non-adjusted and adjusted solid precipitation data are implemented in 20 hydrological models, revealing the impact of the adjustments on hydrological simulation and water balance over a boreal watershed. Hydrometeorological data used for validation comes from the Bassin Expérimental du Ruisseau des Eaux Volées (BEREV) from 2003 to 2018.

[1] Faculté de foresterie, de géographie et de géomatique, Université Laval, Québec City, QC, Canada
[2] Département de genie civil et de génie des eaux, Université Laval, Québec City, QC, Canada

An Improved Technique for Post-Processing Solid Precipitation Time Series from Automated Accumulation Gauges

AMBER ROSS[1], CRAIG D. SMITH[1], AND ALAN BARR[1,2]

ABSTRACT

The unconditioned data retrieved from automated operational accumulating precipitation gauges is inherently noisy due to the sensitivity of the instruments to mechanical and electrical interference. This noise, combined with diurnal oscillations and signal drift from evaporation of the bucket contents, can make accurate precipitation estimates very challenging. Compared with rainfall, relative errors are exacerbated for the measurement of solid precipitation because of lower accumulation rates and the systematic undercatch of solid precipitation due to wind. We have explored three post-processing techniques to filter cumulative precipitation time series derived from high-frequency bucket weight measurements: Operational 15 Minute (O15), Neutral Aggregating Filter (NAF), and Supervised Neutral Aggregating Filter (NAF-S). Inherent biases in these post-processing techniques have uncovered the need for a robust automated method to derive a clean precipitation time series from high-frequency bucket weight measurements that have varying levels of noise, diurnal signals, and evaporation. This study looks at the issues with current post-processing techniques and introduces a new automated method called the Segmented Neutral Aggregating Filter (NAF-SEG). The automated NAF-SEG technique filters 1-minute cumulative precipitation time series in 24-hour segments within three overlapping moving windows per day. The evaluation utilizes simulated data as a control but also applies the technique to real-world data collected from several WMO Solid Precipitation Inter-Comparison Experiment (SPICE) sites. Performance metrics are characterized using total seasonal bias, RMSE, and Pearson's correlation coefficient. The NAF-SEG post-processing technique shows substantial improvement over current automated techniques, contributing to the overall accuracy of gauge measurements of solid precipitation.

---

[1] Environment and Climate Change Canada, Climate Research Division, Saskatoon, SK, Canada
[2] Global Institute for Water Security, University of Saskatchewan, Saskatoon, SK, Canada

The Development and Testing of WMO-SPICE Tipping Bucket Precipitation Gauge Adjustments

JOHN KOCHENDORFER[1], MICHAEL EARL[2], AND DANIEL HODYSS[3]

ABSTRACT

Even though many solid precipitation measurements are recorded using heated tipping bucket gauges, the performance of these gauges for the measurement of solid precipitation has not been well characterized. Tipping-bucket gauges can suffer from significant measurement delays, as precipitation accumulated in the gauge funnel must be melted in sufficient quantity to trigger a full tip before being measured. In addition, underestimates of precipitation may be worse for tipping-bucket snowfall measurements than for weighing gauges, as both evaporation and wind may remove precipitation from a gauge funnel before it can be measured. Five different types of heated tipping-bucket gauges were evaluated at six different sites for the World Meteorological Organization Solid Precipitation InterComparison Experiment (WMO-SPICE). These results were used to develop and evaluate adjustments for the undercatch of solid precipitation. New methods to optimize and test precipitation adjustment transfer functions were developed to address challenges specific to tipping bucket measurements. These new methods may also be applicable to other types of precipitation gauges. The new transfer function development methods were compared to more traditional catch efficiency type methods. In addition, a more general, multi-gauge, multi-site transfer function was developed. This new transfer function is more generally applicable for many types of heated tipping bucket gauge measurements.

[1] NOAA Atmospheric Turbulence and Diffusion Division, Oak Ridge, TN, USA
[2] Meteorological Service of Canada, Environment and Climate Change Canada, Dartmouth, NS, Canada
[3] Naval Research Laboratory, Monterey, CA, USA

Snow Depth and Snow Water Equivalent Data at Stations Included in the GHCN Database

KATHLEEN F. JONES[1] AND STEVEN F. DALY[1]

ABSTRACT

The Global Historical Climatology Network is a worldwide database of daily weather data from over 107,000 surface stations compiled by NOAA's National Centers for Environmental Information. Snow depth is one of five core elements reported. More than 15,000 of these stations also report snow water equivalent (SWE) as the depth of melted snow. We have extracted snow depth and SWE data for the days at stations at which both elements are reported. In this paper, we present statistics of the snow to liquid ratio (SLR) and underlying snow depth and SWE data to a) identify issues with the data and b) characterize the snow pack using SLR quantiles. SLR can be an indicator of problems with the snow depth data, the SWE data, or both. Some ratios occur more frequently than would be expected statistically or physically. This may be due to measurement errors, recording or transcription errors, inconsistent units, or rule-of-thumb estimates reported as observations. We remove the values flagged by GHCN as questionable and compute SLRs for the remaining data, compiled into the Sturm *et al.* (1995) snow classes. Quantiles of SLR in these snow classes, binned by snow depth and day of year, characterize the median density and density variation of the snow on the ground.

Keywords: GHCN, snow depth, snow water equivalent, snow liquid ratio, snow class

INTRODUCTION

The Global Historical Climatology Network (GHCN) is a worldwide database of daily weather data from over 107,000 surface stations compiled by NOAA's National Centers for Environmental Information (Menne *et al.*, 2012a,b). The depth of snow on the ground is one of five core elements reported. More than 15,000 of these stations also report the water equivalent of the snow on the ground, as the depth of the melted snow. Our focus is the ratio of the snow depth to the snow water equivalent, the snow-to-liquid-equivalent ratio (SLR) used in Baxter *et al.* (2005) and Alcott and Steenburgh (2010) to describe the ratio of new snowfall to the depth of the melted water equivalent. Huntington (2005) calculates monthly average SLRs for November through April in New England. Meyer *et al.* (2012) use the SLR climatology for December, January and February to investigate discrepancies between SWE and accumulated precipitation.

In this paper, we use SLR as the ratio of the reported depths of snow on the ground and water equivalent snow on the ground. In a study of snow loads in Alaska (Jones and Daly, 2016) we found that daily SLR reported at many airport weather stations in Alaska is exactly 10 relatively frequently, which we attributed to the reported SWE being calculated from the snow depth rather than measured. In this paper, we examine the distribution of SLR for the networks providing snow data

---

[1] U.S. Army Corps of Engineers Cold Regions Research and Engineering Laboratory, Hanover, NH, USA

to GHCN, using all the available pairs of daily measurements in the database. We also compile the data into snow classes and use SLR quantiles to determine typical values and the range of values. A comparison of the GHCN SLR quantiles with the SLR determined from the Sturm *et al.* (2010) (S2010 in the following) bulk densities is also presented.

In the station files in the GHCN database, the snow depth and SWE elements are identified as SNWD and WESD, respectively. SNWD is reported in whole millimeters and WESD in 0.1-mm increments. Simultaneous pairs of these elements have been reported at 13,772 stations in the database at least once. Most of these stations are in one of four networks. In Section 2, we provide information on each these networks. Cumulative distributions and probabilities for the network SLRs are discussed in Section 3. In Section 4, we present and discuss the snow class SLR statistics, using the Sturm *et al.* (1995) snow classes.

NETWORKS

Station names in the GHCN database begin with the country code followed by a one-character network code. The networks with SWE and snow depth data have the following network codes (*ftp://ftp.ncdc.noaa.gov/pub/data/ghcn/daily/readme.txt*):

- 0: unspecified (one station in Canada)
- 1: Community Collaborative Rain, Hail, and Snow (CoCoRaHS) (11,117 stations in the US and 373 in Canada)
- C: U.S. Cooperative Network (COOP) (942 stations)
- M: World Meteorological Organization (one station in Antarctica)
- S: U.S. Natural Resources Conservation Service SNOwpack TELemtry (SNOTEL) (820 stations)
- W: Weather Bureau Army Navy (WBAN) (435 stations in the US and 83 stations in Austria, Antarctica, Canada, France, Greenland, Iceland, Italy, Japan, North Korea, South Korea, or United Kingdom)

The station in Canada with an unspecified network is at Stephenville Airport in Newfoundland and has depth and SWE measurements between 1954 and 1958. The station in Antarctica in the WMO network is at McMurdo Sound NAF (Naval Air Facility). Snow depth and SWE measurements were made there between 1958 and 1964. Based on those station characteristics we group both stations with the WBAN network for this paper.

We describe the characteristics of the snow reports from the CoCoRaHS, COOP, SNOTEL, and WBAN networks in the following.

CoCoRaHS

The Community Collaborative Rain, Hail, and Snow Network was started at Colorado State University's Colorado Climate Center in 1998. The network is sponsored by NOAA and the National Science Foundation. Observers are volunteers. The network spread to other states beginning in 2003 and as of 2013 included locations in all 50 states and the District of Columbia (*https://www.cocorahs.org/Media/Docs/StateAdmissiontotheCoCorahsUnion2018.pdf*). Locations in Canada were added beginning in 2012. Observers purchase the CoCoRaHS 4-inch-diameter high-capacity manual rain gauge and measure daily precipitation at 7 am. The CoCoRaHS web site has training slide shows on measuring snow depth and snow water equivalent (*https://www.cocorahs.org/Content.aspx?page=training_slideshows*). At the observation time, observers use a snow ruler to measure the snow depth on the ground and record the average depth to the nearest 0.5 inch. If there is snow on more than half of the area, the snow depth is reported as the average of the bare and snow-covered depths. If more than half of the ground is bare, Trace is reported. Reporting the water equivalent of total snow on the ground is optional. Observers who choose to measure this parameter are asked to make that measurement once a week, with Monday suggested as a good day. Observers are instructed to take a snow sample from an area with the average snow depth. They use their 4-inch rain gauge to collect the snow core. In deeper and/or denser snow, instructions for making and using a core tube, made from a section of 2-inch ABS black plastic drain pipe, with teeth filed in one end and a cap for the other end, are provided. Observers melt the snow core and measure the depth of the water to the nearest 0.01 inch using the inner tube of the precipitation gauge. Instructions are also provided for measuring the snow water equivalent by weight using a digital balance, such as a commercial nutrition scale. The document *https://www.cocorahs.org/Media/Docs/CoCoRaHS_QA_QC_Nov_2018.pdf* describes quality control of the data. The automated checks on the web-based data entry forms require that reported SWE is not more than 90% of the snow depth. Manual QC includes creating national maps and checking to identify very high values of total snow depth that seem erroneous. They find that most errors are reporting rather than measuring problems, including misplaced decimal points.

COOP

In the National Weather Service's Cooperative Observers Program, volunteers or contractors in the United States make daily observations (*https://www.weather.gov/coop/Overview*). At COOP sites that are collocated with standard observing stations, the COOP data is documented independently. The equipment at COOP sites must meet NWS standards and may be owned by the NWS, the observer, or some other entity. COOP observers making snow measurements are referred to NWS (2014) for the procedures for measuring the depth of snow on the ground and the snow water equivalent. Snow depth is measured with a measuring stick, sometimes by averaging a number of depths within 300 ft of the official observing location, excluding paved areas or those disturbed by human activity. At locations where snow tends to be deep, snow stakes, ideally located on level ground in an area without trees, buildings, and other obstructions may be used instead. Observers are to use their good judgement in determining the average snow depth in an area where it varies markedly. If less than 50% of the ground is covered by snow, Trace should be reported. Where drifting has occurred, the observer averages measurements taken in areas least affected by drifting. Depths are reported to the nearest whole inch, rounding up at 0.5 inch and greater. For measuring the water content of the snow, using the outer cylinder of a 4-inch rain gauge to get a snow core is suggested. The water equivalent is measured either by allowing the snow in the cylinder to melt, or melting the snow by adding a measured amount of warm water, which is subtracted after the liquid measurement is made. In either case, the liquid that results is poured through the funnel into the inner measuring tube of the rain gauge. The water content is to be reported "...in accordance with the directions provided by the observing program...". In NWS (2017) in Appendix B COOP Observer - Observation Instructions, *Section 2 Precipitation* says that snow depths are recorded in whole inches and the water content of the snow on the ground is reported in hundredths of an inch. In *Section 7 River Stage Observations*, observers who are instructed to measure the water content of snow or ice are to report to the nearest 0.1 inch.

SNOTEL

The SNOTEL network currently consists of 867 automated sites at remote high elevations in the western U.S. states including Alaska. The sites are powered by batteries charged by solar cells and are intended to operate unattended without maintenance for at least a year. Measurements are made hourly and stored on a data logger in the equipment shelter at the site. Data transmission from the sites to the National Water and Climate Center (NWCC) is by telemetry. The site instrumentation includes a sonic snow depth sensor and a snow pillow instrumented with a pressure transducer. The snow depth sensor has 0.5-inch resolution and an accuracy of ± 2 inches or 0.4% of the distance to the snow surface. The snow water equivalent is calculated from the snow pillow measurements with a 0.1-inch resolution up to 250 inches of water and an accuracy of ± 4% (*https://directives.sc.egov.usda.gov/OpenNonWebContent.aspx?content=27630.wba*).

WBAN

NOAA (2017) requires snow depth reports at designated stations at 0000, 0600, 1200, and 1800 UTC, whenever there is more than a trace of snow on the ground. Snow depth is reported using three digits in whole inches in group 4 after RMK (remark) in a METAR (Meteorological Terminal

Aviation Routine Weather Report). For example, 5 inches of snow is coded as 4/005. The water equivalent of snow on the ground is reported at designated stations at 1800 UTC if the average snow depth is at least 2 inches. The value is recorded to 0.1 inch using three digits beginning with 933 after RMK. For example, a SWE of 1.2 inches is coded as 933012. SWE is never estimated, assuming, for example, a 10 to 1 ratio or temperature/snow water equivalency tables.

Comparison

Differences in the locations of these stations and in the way the snow measurements are made result in differences in the distribution of SLR for these networks. Here we provide general information about the station data. In plots of this network data, we identify CoCORaHS by nw1, COOP by nwC, SNOTEL by nwS and WBAN by nwW.

Figure 1 shows the distribution of the elevation of stations in each of the networks. The SNOTEL stations are the highest, with a median elevation of 2366 m. Most of the CoCoRAHS stations are at relatively low elevations (median=289 m), but 24% of stations are at elevations greater than 1000 m. The COOP stations are all at low elevations, with a median of 281 m. The WBAN stations also tend to be at low elevations, with a median at 199 m, but 12% are above 1000 m.

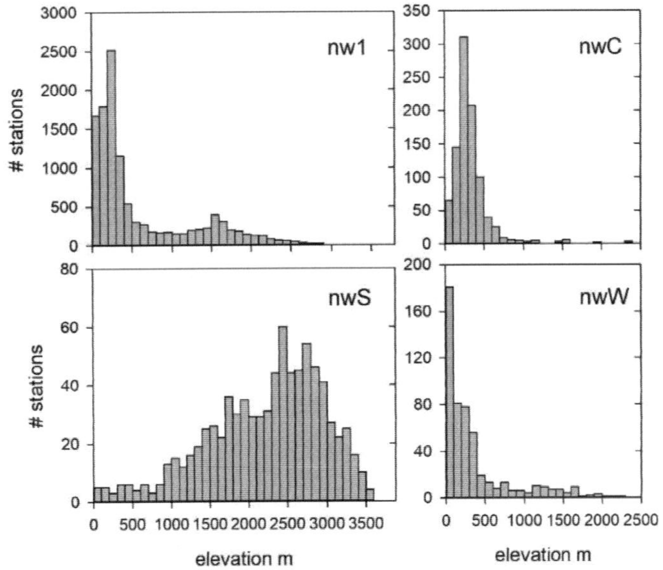

Figure 1. Station elevation distribution for each of the networks (nw1= CoCoRaHS, nwC=COOP, nwS=SNOTEL, nwW=WBAN).

Figures 2 and 3 show the number of snow-SWE pairs in each year and month, respectively, for each network. Most of the measurements in the CoCoRaHS network and the vast majority of those in the COOP network are since 2005. Snow depth and SWE pairs began being reported in the SNOTEL network in the early 1980s, with most reports after 1995. In contrast, the WBAN report counts began decreasing in the 1990s, perhaps in conjunction with the change from human observers at most of the first-order weather stations in the United States to ASOS (Automatic Service Observing System). After three years with no snow reports at all, a few WBAN stations resumed reporting snow data in 2009. The differences between networks in the distribution of the number of reports in each month (Figure 3) is consistent with the higher elevations of the SNOTEL stations, with many reports in May and June.

Frequency of reporting may vary by day of week (Figure 4). CoCoRaHS observers who measure SWE are asked to do that once a week, preferably on Monday. The reports on other days may indicate observers who choose a different day for their weekly reports, or observers who measure more frequently. COOP observers also tend to make their observations on Mondays. SNOTEL and WBAN observations are daily.

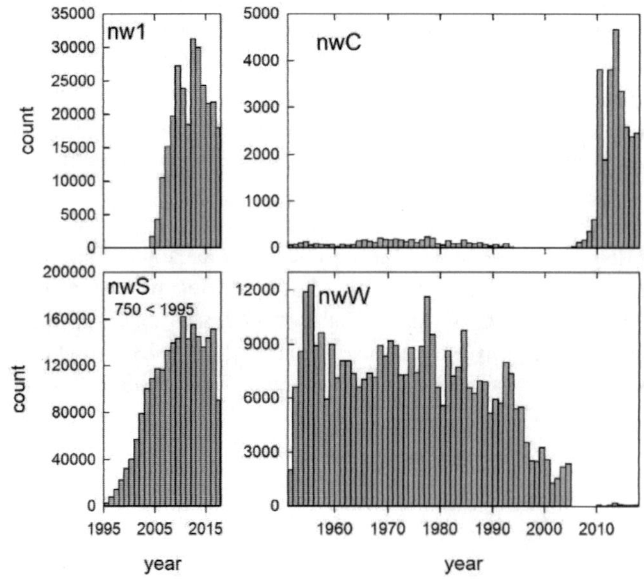

Figure 2. Number of snow depth-SWE data pairs by year for each network. In the SNOTEL network, 750 data pairs between 1983 and 1994 are not shown.

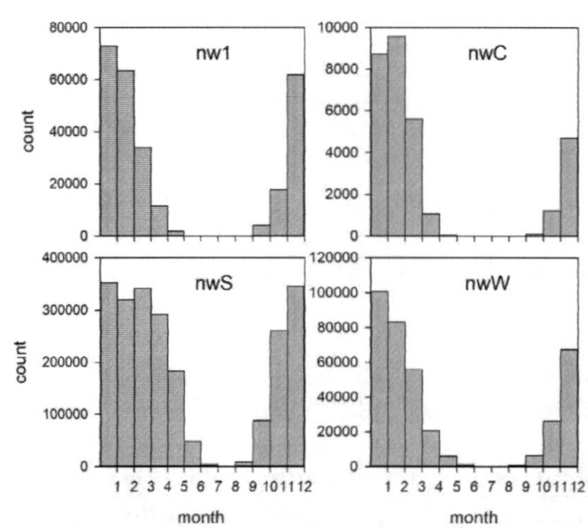

Figure 3. Number of snow depth-SWE pairs by month for each network.

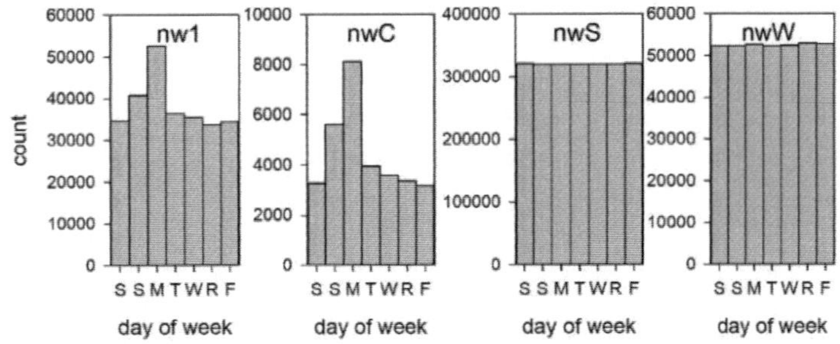

Figure 4. Number of snow depth-SWE pairs by day of week for each network. Observers in the CoCoRaHS network are asked to measure SWE once a week.

Figures 5 and 6 show the distributions of snow depth and SWE, respectively, for these networks. The SNOTEL network is clearly different from the others, with frequent occurrences of large snow depths and large SWE. The CoCoRaHS stations have somewhat smaller depth and SWE values than the stations in the COOP and WBAN networks. We can characterize the snow in these networks by the 10th and 90th percentile snow depth and SWE values, shown in Table 1. The deep, dense snow at the SNOTEL stations stands out.

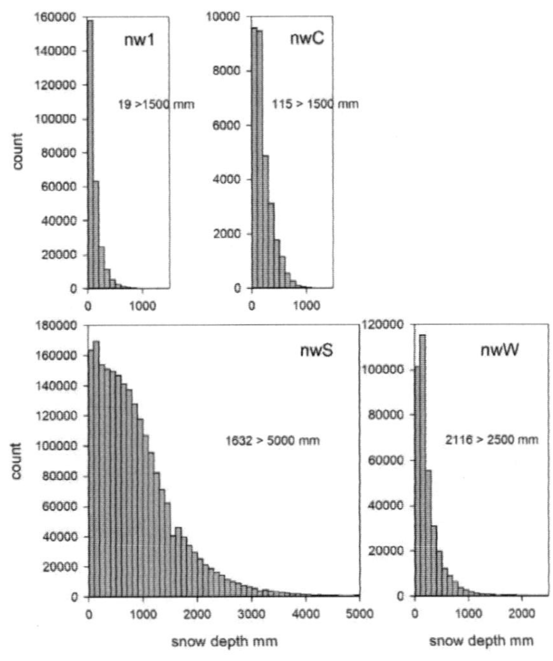

Figure 5. Snow depth frequencies for each network.

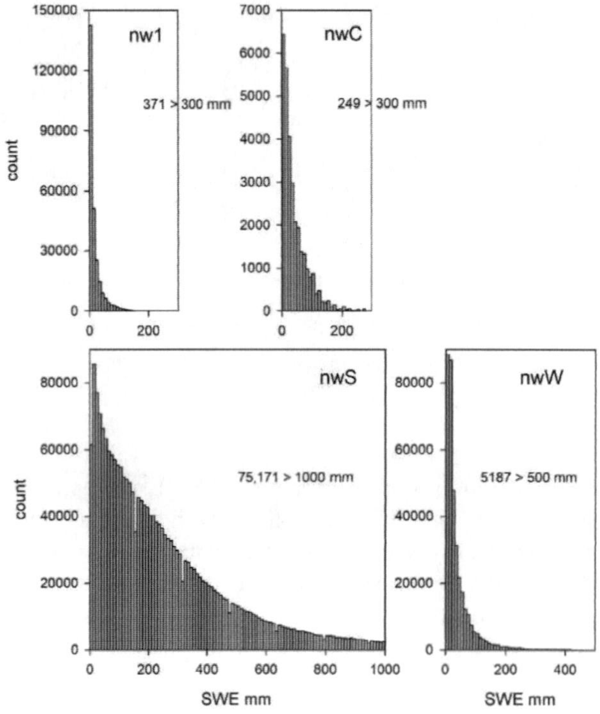

Figure 6. SWE frequencies for each network.

Table 1. 10th and 90th percentile snow depth and SWE for each of the networks.

Network	Snow depth (mm)		SWE (mm)	
	10th%	90th%	10th%	90th%
CoCoRaHS	13	264	1.0	47.2
COOP	46	419	4.8	101.6
SNOTEL	102	1829	27.9	645.2
WBAN	51	559	5.1	101.6

The precision of the snow depth and SWE reports from the stations in each network can be shown by the probabilities of each 1 mm increment in snow depth and each 0.1-mm increment in SWE. These probabilities are plotted in Figures 7 and 8, respectively, for snow depths up to 15 inches and SWE up to 1.5 inches. Figure 7 shows that snow depths are sometimes reported to 0.1 inch even though observers are asked to report in 0.5- or 1-inch increments. The only network that never violates their 1-inch rule is SNOTEL. Observers in both the CoCoRaHS and COOP networks are asked to report SWE in 0.01-inch increments. While CoCoRaHS observers do that, most COOP observations are in 0.1-inch increments (Figure 8). In contrast, many WBAN reports are in 0.01-inch increments, even though the required METAR report only allows for 0.1-inch increments. Those reports must predate this METAR format. We list the snow depth and SWE increments commonly used in reports for each network in Table 2. The snow to liquid ratio SLR is the ratio of these simultaneous reports of snow depth and SWE at a station. In the next section, we look at the distribution of SLR values for these networks.

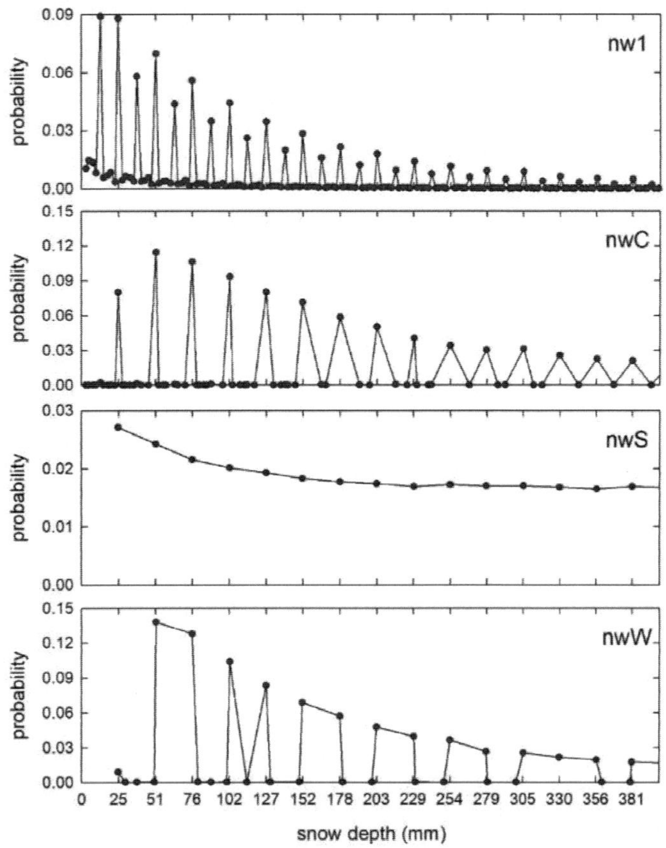

Figure 7. Probability of occurrence of reported snow depth in 1-mm increments for each network. Ticks along the x-axis are in 1-inch increments. Snow depth are reported in whole inches most of the time, except for the CoCoRaHS network where half-inch increments are commonly used.

Table 2. Typical snow depth and SWE report increments

Network	Snow depth (in.)	SWE (in.)
CoCoRaHS	0.5	0.01
COOP	1	0.1
SNOTEL	1	0.1
WBAN	1	0.1

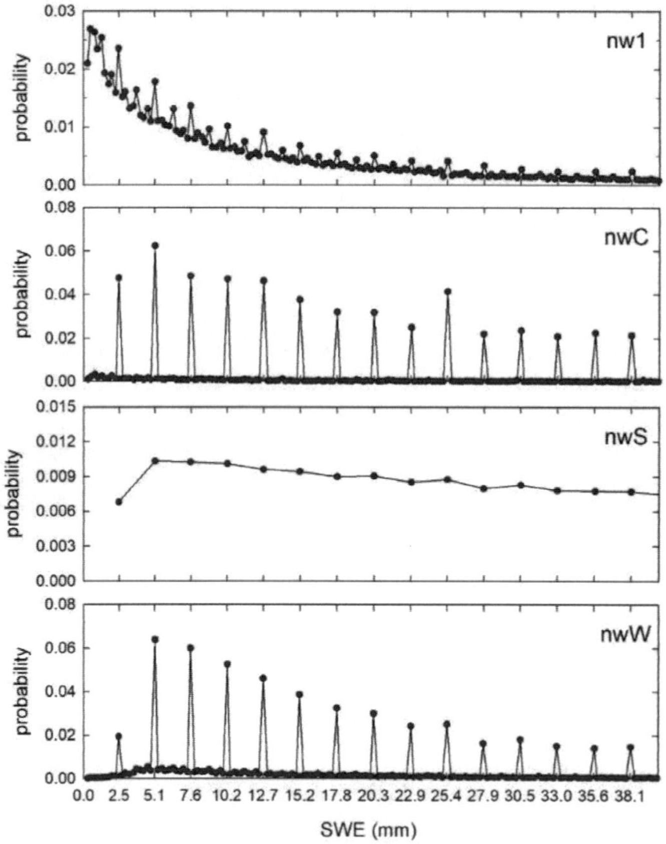

Figure 8. Probability of occurrence of SWE in 0.1-mm increments for each network. Ticks along the x-axis are at 0.1-inch increments. SWE is reported in 0.1-inch increments most of the time, except for the CoCoRaHS network where 0.01-inch increments are commonly used.

SNOW TO LIQUID RATIO

Cumulative distribution

For each snow depth-SWE pair, we calculated SLR and, for each network, ordered the N SLR from smallest to largest and generated the cumulative distribution of SLR (Figure 9) by assigning the probability of non-exceedance to the ordered SLR beginning with $1/(N+1)$. The physical lower bound for SLR is 1.09, which would only occur if the snow were bubble-free ice with a density of 0.92 g cm^{-3}. It is likely that many of the snow depth-SWE pairs that result in SLRs slightly larger than 1.09 are also erroneous. Values smaller than 1.09 occur most frequently in the COOP network In each network, there is a significant fraction of SLR exactly equal to 10, indicated by the length of the vertical line along the gridline at 10. The occurrence rates of SLR < 1.09 and SLR = 10 for all four networks are shown in Table 3.

Table 3. Occurrence rates of SLR anomalies by network. SLR < 1.09 (first column) is a physical anomaly. Too frequent occurrences of SLR exactly equal to 10 is a statistical anomaly.

Network	SLR < 1.09	SLR = 10	Synthetic SLR = 10
CoCoRaHS	1.1%	5.6%	1.0%
COOP	6.8%	11.2%	5.2%
SNOTEL	0.3%	1.4%	0.9%
WBAN	0.3%	16.9%	5.3%

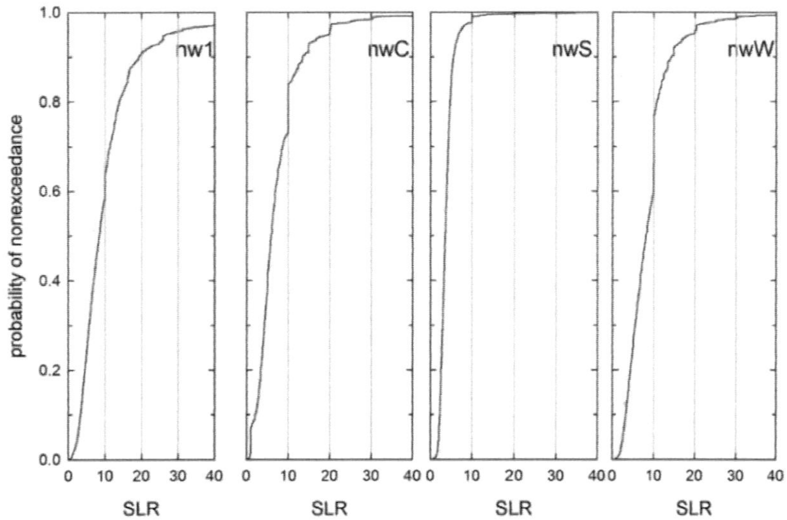

Figure 9. Probability of exceedance plots for SLR for each of the GHCN network with snow depth and SWE data. A long vertical line at SLR = 10 may indicate that the reported SWE is frequently calculated from the snow depth by dividing by 10, rather than measured.

Small SLR

The frequency of SLR < 1.09 varies by year. It is relatively high from 1983-1993 in the SNOTEL network, when there were only 504 reports altogether. During those years, the SNOTEL QC effort was apparently spinning up, as there are relatively few small SLR subsequently. In the COOP and WBAN networks, the frequency of small SLRs is relatively high from 2011 to the present. Those years are characterized by a dramatic increase in simultaneous COOP reports of snow depth and SWE and an increase in WBAN reports following three years of no snow depth-SWE reports. At the WBAN stations, this change may be associated with replacing human observers by ASOS for most weather elements and then ultimately restarting snow measurements at some stations. The current QC program at COOP and WBAN stations is apparently not catching these low SLR discrepancies as well as it did in the past. These anomalous SLRs may result from measurement errors or from misplaced decimal points or missing or extra leading zeros in recording or transcribing the data.

If snow depth reports are too low by a factor of ten and/or SWE reports are too high by a factor of ten, associated with recording or transcription errors, we would expect many of these reports to be outliers. Table 4 lists the fractions of the snow and SWE values resulting in SLR < 1.09 that are smaller than the 10th% values or greater than the 90th% values for each network. Note that in the SNOTEL network small SLR values are typically associated with very small snow depths. In the other three networks, very small snow depths and very large SWE are about equally likely. By careful examination of the snow depth and SWE time series at a station before and after these anomalous SLRs, one might be able to correct many of these outliers by moving the decimal point. That is beyond the scope of this paper.

Table 4. Fraction of the snow depth and SWE values resulting in SLR < 1.09 that are below the 10th and above the 90th % values

Network	Snow depth		SWE	
	<10th%	>90th%	<10th%	>90th%
CoCoRaHS	47%	2%	0	38%
COOP	24%	4%	0	47%
SNOTEL	77%	0	16%	5%
WBAN	42%	1%	0	48%

SLR = 10

With snow depth in whole mm and SWE in 10[th]s of mm, SLR values of 10.000 occur only when the snow depth is exactly 10 times SWE up to snow depths of 20000 mm (20 m). That is, there are no values, like 10.0001 or 9.9996, that would round to 10.000. The largest snow depth of 15240 mm (WBAN) in our compilation is less than 20000, so SLR = 10.000 occurs only when the snow depth is exactly 10 times SWE. We compared the occurrence rates in Table 3 with expected occurrence rates calculated for synthetic distributions of snow depth and SWE obtained by fitting to their cumulative distributions between the 10th% and 90th% values (Table 1). That distribution provides the relative probabilities for the increments in Table 2. We then calculated SLR for all snow depth and SWE values in that range weighted by their probability. The expected rates for the four networks are in the third column of Table 3. The actual occurrence rates for exact 10 SLRs is substantially higher than these expected rates in the WBAN network, and, to a lesser extent, the COOP and CoCoRaHS networks.

Quality Assurance

The GHCN database includes a quality assurance flag for each snow depth and SWE report. Overall, 0.7% of the snow depth-SWE pairs fail the QA check, with the fraction varying from 0.3% for the WBAN network to 0.7% for the SNOTEL network. Snow depth QA failures are ten times more frequent than SWE failures, with the three most common failure reasons being 1) gap check, 2) bounds check, and 3) internal consistency check. The most common reason for a SWE report to fail the QA check is "failed temporal consistency". There is no indication the QA checks include a comparison of the snow depth and SWE measurements: fewer than 2% of the QA failures would result in SLR<1.09 and only 3% of the failures would produce SWE = 10.000. The most common reason a record with SLR = 10.000 fails is "failed streak/frequency check". Those records are almost all from Point Barrow, Alaska, when 6 inches of snow was reported every day from 19 January to 25 April 1974 and 4 inches was reported from 20 December 1975 to 4 May 1976.

SNOW CLASS AND SLR

We compiled the snow-SWE pairs into their Sturm *et al.* (1995) snow class, using nl_sclass (*https://arcticdata.io/catalog/view/doi:10.5065/D69G5JX5*), which has 25 km rasters. Of the 13,772 GHCN stations with simultaneous snow depth and SWE reports, 641 are in the nl_sclass Water class because of the rough definition of the coastline by the 25 km rasters. We assigned those stations to the snow class of the closest land raster, most of them to Ephemeral (310) or Maritime (289). Note that other GHCN stations may also not be in to the correct snow class because of the rough boundary definition. This problem may be most pronounced for the Alpine region because of its large perimeter to area ratio, resulting in some stations that should be designated as Alpine falling in the surrounding Prairie rasters. Table 5 shows the snow classes with the total number of GHCN snow depth-SWE reports without a QA flag and the fraction of reports in that snow class from each network. The snow class with the most reports is Prairie, followed by Alpine and Maritime. The Taiga snow class has the fewest cases. The SNOTEL network is the source of the majority of cases in every snow class except Tundra. The majority of the Tundra data is from WBAN and there is no contribution from COOP. Figure 10 shows the distribution of stations, identified by network, for the North American extent of each snow class.

We examined the statistics of SLR for this GHCN data set for each snow class to determine typical snow pack characteristics and quantify the variation. We also compare the GHCN SLR quantiles to SLR from the S2010 fitted bulk snow density based on day of year *DOY* and snow depth *snd* [cm]:

$$\rho_b[\text{g cm}^{-3}] = (\rho_{max} - \rho_0)(1 - e^{-k_1 snd - k_2 DOY}) + \rho_0 \quad (1)$$

Values for parameters ρ_{max}, ρ_0, k_1, and k_2 depend on the snow class and are listed in Table 6 (S2010). There are no parameter values for the Ephemeral snow class.

Table 5. Contribution of the four networks to each of the snow classes

Snow class	Tundra	Taiga	Maritime	Ephemeral	Prairie	Alpine
# cases	61,027	36,427	772,633	115,262	1,003,883	872,505
Network						
CoCoRaHS	0.1%	0.7%	15.2%	25.8%	8.3%	4.0%
COOP	0.0%	2.4%	2.0%	1.3%	0.8%	0.6%
SNOTEL	7.7%	57.5%	62.2%	62.4%	81.7%	92.7%
WBAN	92.2%	39.4%	20.7%	10.5%	9.2%	2.7%

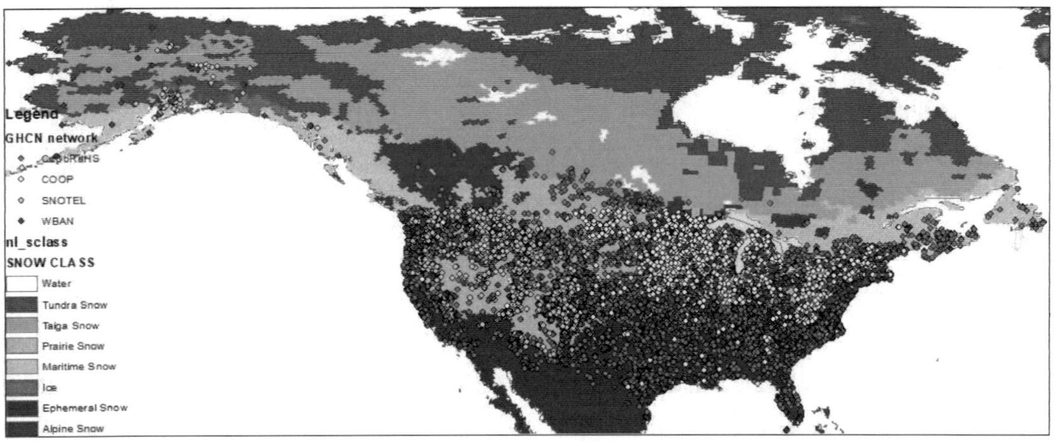

Figure 10. Stations in the GHCN database, identified by network, with snow depth and SWE data, mapped on the Sturm *et al.* (1995) snow classes in North America

Table 6. Parameter values for (1) for each snow class.

Snow Class	ρ_{max}	ρ_0	k_1	k_2
Tundra	0.3630	0.2425	0.0029	0.0049
Taiga	0.2170	0.2170	0	0
Maritime	0.5979	0.2578	0.0010	0.0038
Prairie	0.5940	0.2332	0.0016	0.0031
Alpine	0.5975	0.2237	0.0012	0.0038

To compute the GHCN SLR statistics and compare to S2010, we separated the data into 1-inch increments (0.025 m) of snow depth and 2-day increments of *DOY*. This two-day grouping somewhat ameliorates the less-than-daily measurements of SWE at CoCoRaHS and COOP stations, making it more likely that the bins have enough data to calculate useful quantiles. We assigned data to a day-depth bin by rounding the measured depth to the nearest inch and assigning even Julian days to the next-earlier Julian day. This conveniently moves leap day (366) to 365. As used in the S2010 model, *DOY* is calculated by subtracting 366 from Julian day 274 (1 October) or greater. Thus it begins at *DOY*= -92 and then continues from -1 to 1 (1 January) and up to 181 (30 June). In each day-depth bin of each snow class, we counted the number of cases. If there are at least ten, we sorted from smallest to largest SLR and determined the 10th, 50th (median), and 90th percentile values.

We plotted those SLR quantiles for a) snow depths up to 3 m at eight *DOY* slices in ~25 day increments from -49 to 125 and b) *DOY* from -92 to 181 at eight snow depth slices ranging from 0.13 to 1.52 m (5, 10, 15, 20, 30, 40, 50, and 60 inches). SLR is plotted on a log scale so that if a 90th% SLR is say, twice, the median and the 10th% SLR for that bin is half the median, the distance to the median in the plot is the same for both. The twice and half SLR ratios in this example correspond to half and twice density ratios.

SLR quantiles and the S2010 fit from (1) for each snow class are in Figures 11a through f, in the same order as in Table 6. Note that there is no S2010 fit for the Ephemeral snow class in Figure 11d. For the GHCN stations, the Tundra and Taiga snow classes (Fig 11a and b) are similar and the Maritime, Prairie, and Alpine classes (Fig 11c, e, and f) are similar.

In the Tundra and Taiga snow classes, snow depths at GHCN stations are shallow, similar to the limited depth range for the data in the S2010 training dataset. The median number of cases in the *DOY*-snow depth bins is 21 for Tundra and 16 for Taiga. The 10th% GHCN SLR tends to be near the S2010 fit for both snow classes, indicating that most of the reported values result in higher SLRs (lower snow densities) than (1) provides. Many of the median Tundra SLRs are 10, so it is likely that SWE was often calculated rather than measured at those WBAN stations.

The GHCN data in the Ephemeral snow class represent somewhat deeper snow than the data in the Tundra and Taiga classes (Fig 11d). That occurs even though the Ephemeral snow season is shorter than for those snow classes, starting later and ending earlier. The median number of cases in the *DOY*-snow depth bins for this class is 19. As the snow depth increases, the 10th and 90th percentile Ephemeral SLRs move toward the median, indicating less variation in the density of deeper snow.

GHCN data in the Maritime, Prairie, and Alpine snow classes represent relatively long snow seasons and deep snow. The median number of cases in each *DOY*-snow depth bin (up to 3 m) is 35, 79, and 66, respectively. For snow depths greater than about 0.5 m (Prairie and Alpine) or 0.75 m (Maritime) the 10th and 90th% SLRs are close to the median values. As the snow gets shallower, the spread between these quantiles increases, indicating increasing variation in snow density. For these three snow classes, all of the SLR quantiles for shallow snow tend to increase in the middle of the snow season and then decrease again at the end of the season, consistent with regularly occurring snowstorms adding relatively low-density snow to the snowpack. This midwinter SLR increase (density decrease) is apparent even for deep snow, particularly for the Alpine snow class and, to a lesser extent, for Prairie snow.

For snow depths over 0.75 m, the GHCN Maritime median SLR (Fig 11c) appears to agree well with S2010. Over that depth range, the median of the 90th% to 50th% ratio is 1.22 and the median of the 10th% to 50th% ratio is 0.81. The ratio of the GHCN 50th% SLR to S2010 is 1.03. Expressed in terms of snow density, 80% of the samples in this class from snow depths of at least 0.75 m have densities within about 20% of the median, and that median is about 97% of the S2010 density fit. These statistics are compiled in Table 7 for the Maritime, Prairie, and Alpine snow classes—the classes that have deep snow at stations in the both the GHCN database and the S2010 test dataset. The median ratios of the 10th and 90th% SLRs to the 50th% value vary only slightly among these snow classes. The median ratios of the GHCN 50th% SLR to the S2010 fit for the Prairie and Alpine snow classes, however, are significantly higher than for the Maritime data, corresponding to snow densities less than 90% of the S2010 density.

For the shallower snow in the Tundra, Taiga, and Ephemeral classes, the 90th% and 10th% SLRs are significantly different from the 50th% SLRs. There is no trend of these quantiles with *DOY* or depth (Fig 11a, b, d) except for a slight trend with *DOY* for Ephemeral snow. Therefore, in Table 8 we show the medians of all three quantiles. The frequent occurrence of SLR = 10 in the Tundra snow class, likely related to the large fraction of cases from WBAN stations, has a significant effect on the calculated quantiles; the 50th% SLR for the snow depth and *DOY* bins is often 10. Therefore, the Tundra SLRs in Table 8 probably do not reflect reality. There is, however, a much smaller contribution of WBAN station data to the Taiga class, and smaller still to the Ephemeral

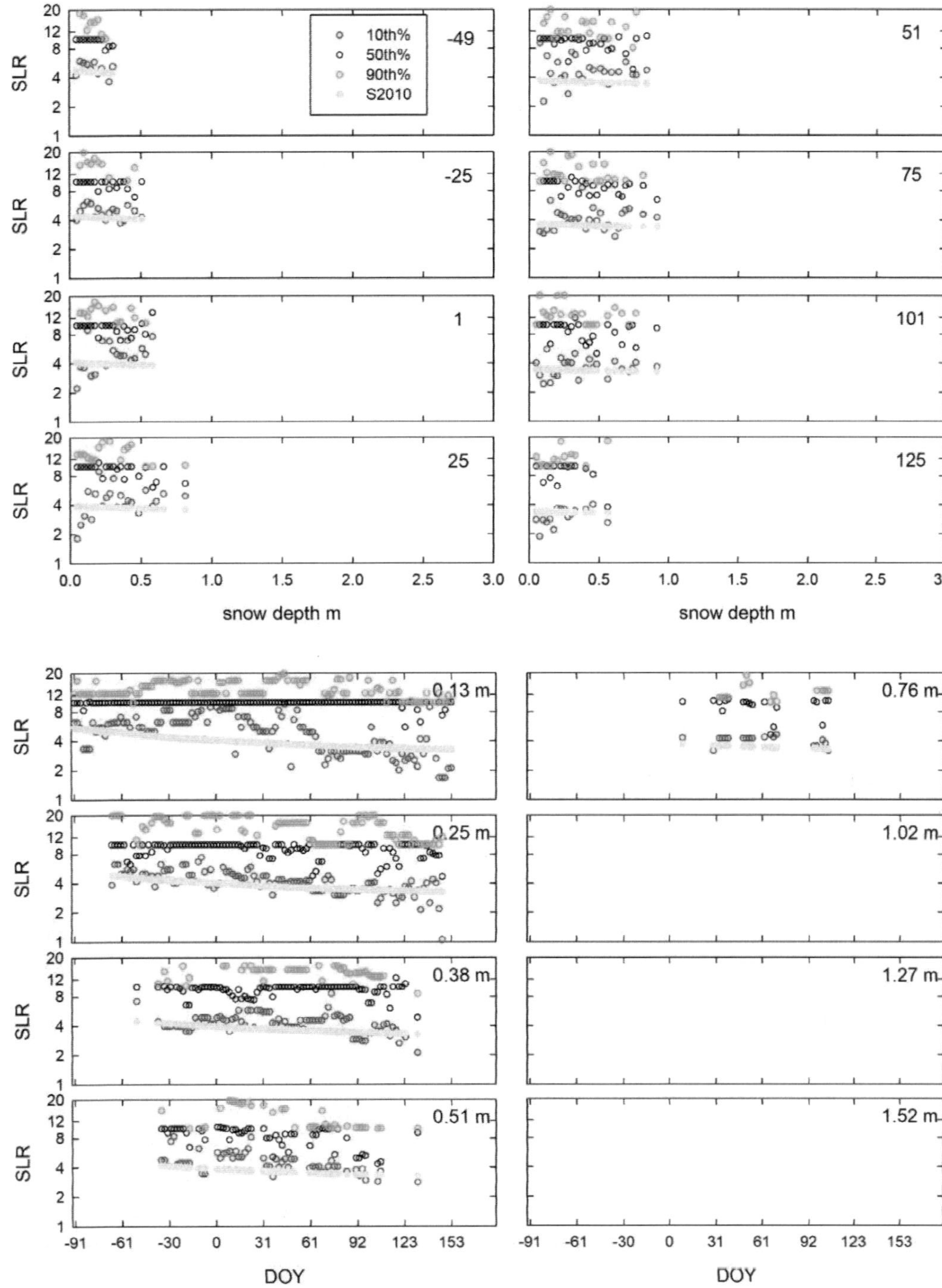

Figure 11a. SLR quantiles for GHCN data in the Tundra snow class: (top) as a function of snow depth for eight *DOY*, (bottom) as a function of *DOY* for eight snow depths. SLRs calculated from the S2010 fit are also shown.

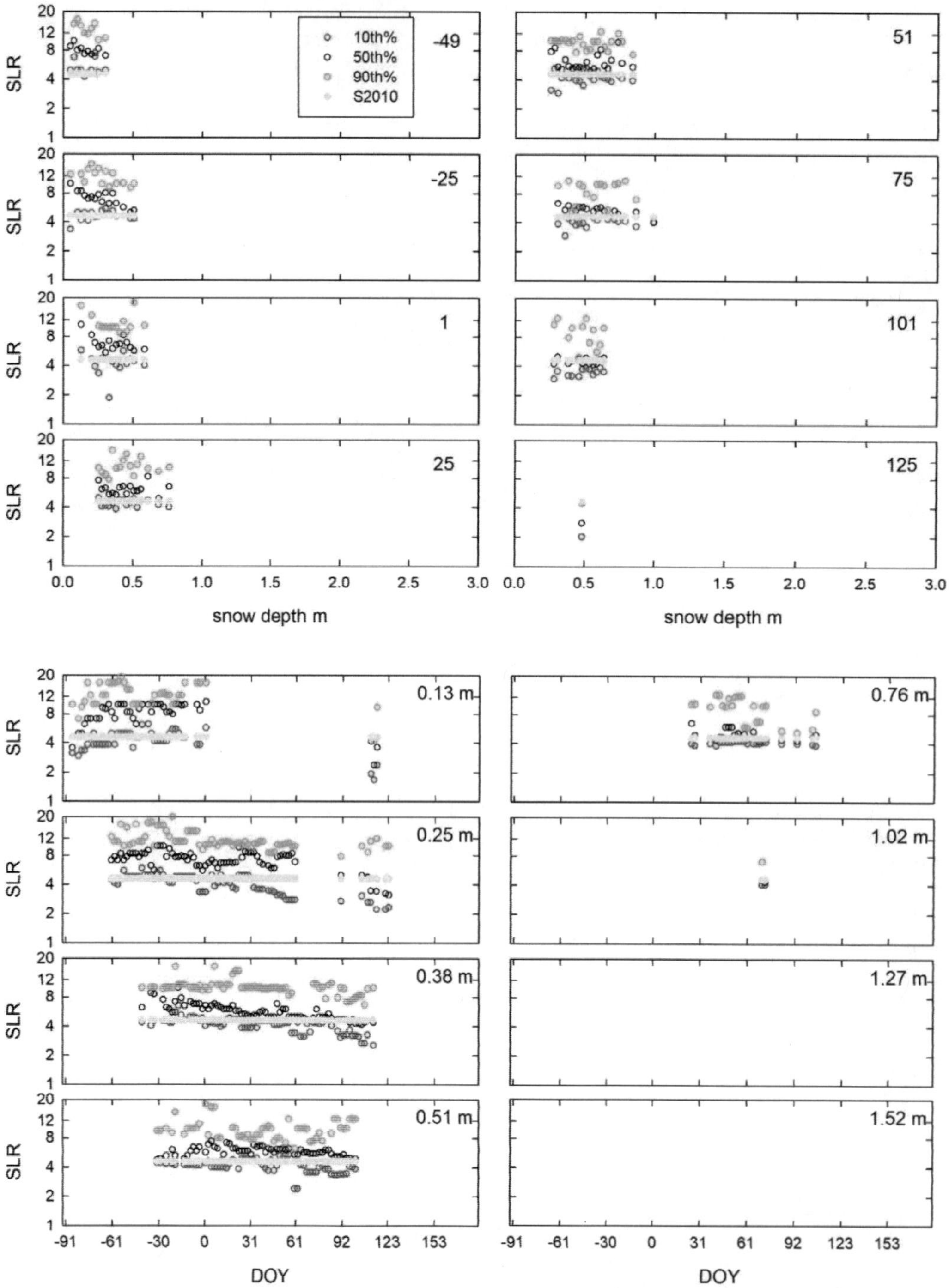

Figure 11b. SLR quantiles for GHCN data in the Taiga snow class: (top) as a function of snow depth for eight *DOY*, (bottom) as a function of *DOY* for eight snow depths. SLRs calculated from the S2010 fit are also shown.

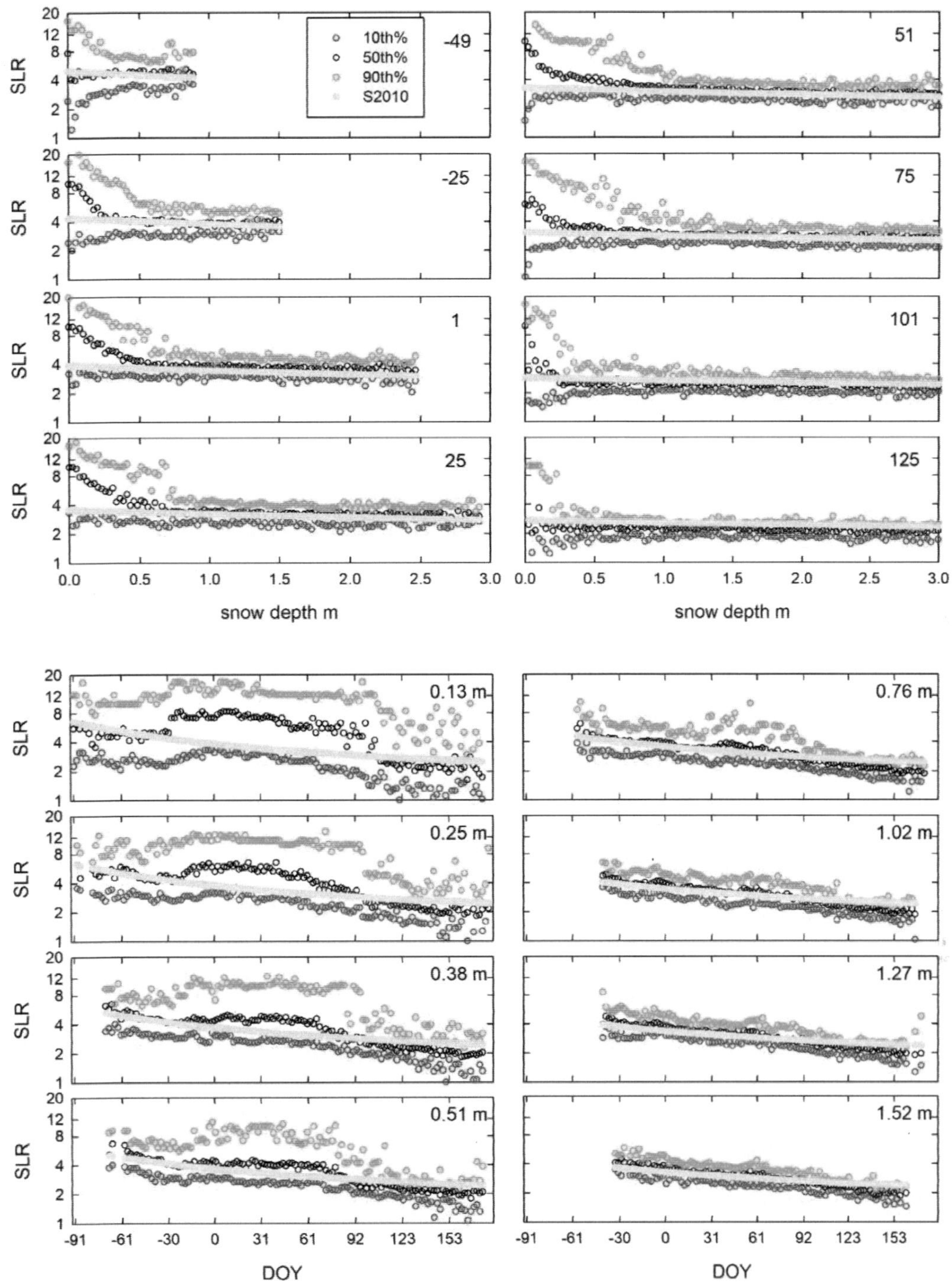

Figure 11c. SLR quantiles for GHCN data in the Maritime snow class: (top) as a function of snow depth for eight *DOY*, (bottom) as a function of *DOY* for eight snow depths. SLRs calculated from the S2010 fit are also shown.

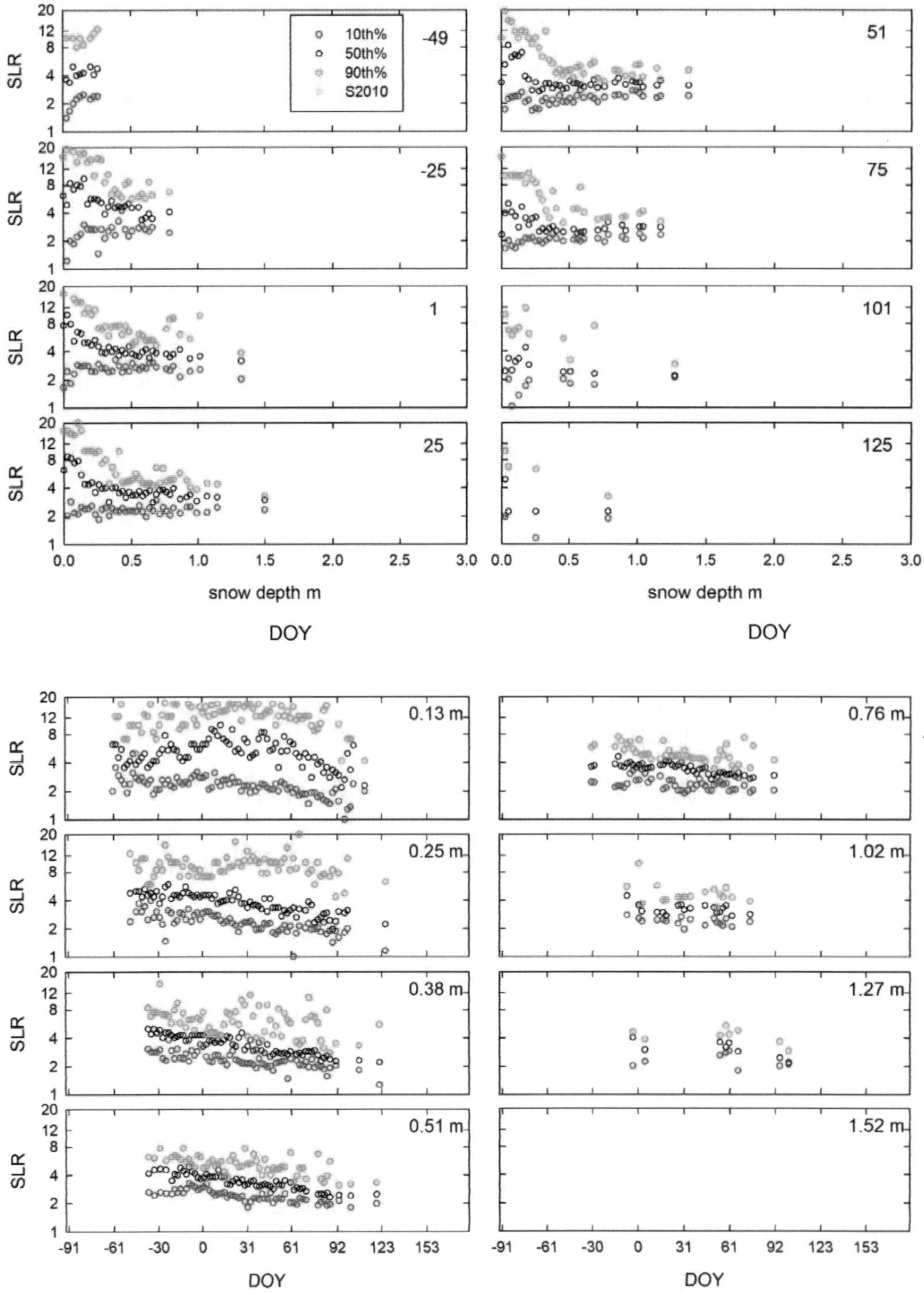

Figure 11d. SLR quantiles for GHCN data in the Ephemeral snow class: (top) as a function of snow depth for eight *DOY*, (bottom) as a function of *DOY* for eight snow depths. S2010 did not determine a fit for this snow class.

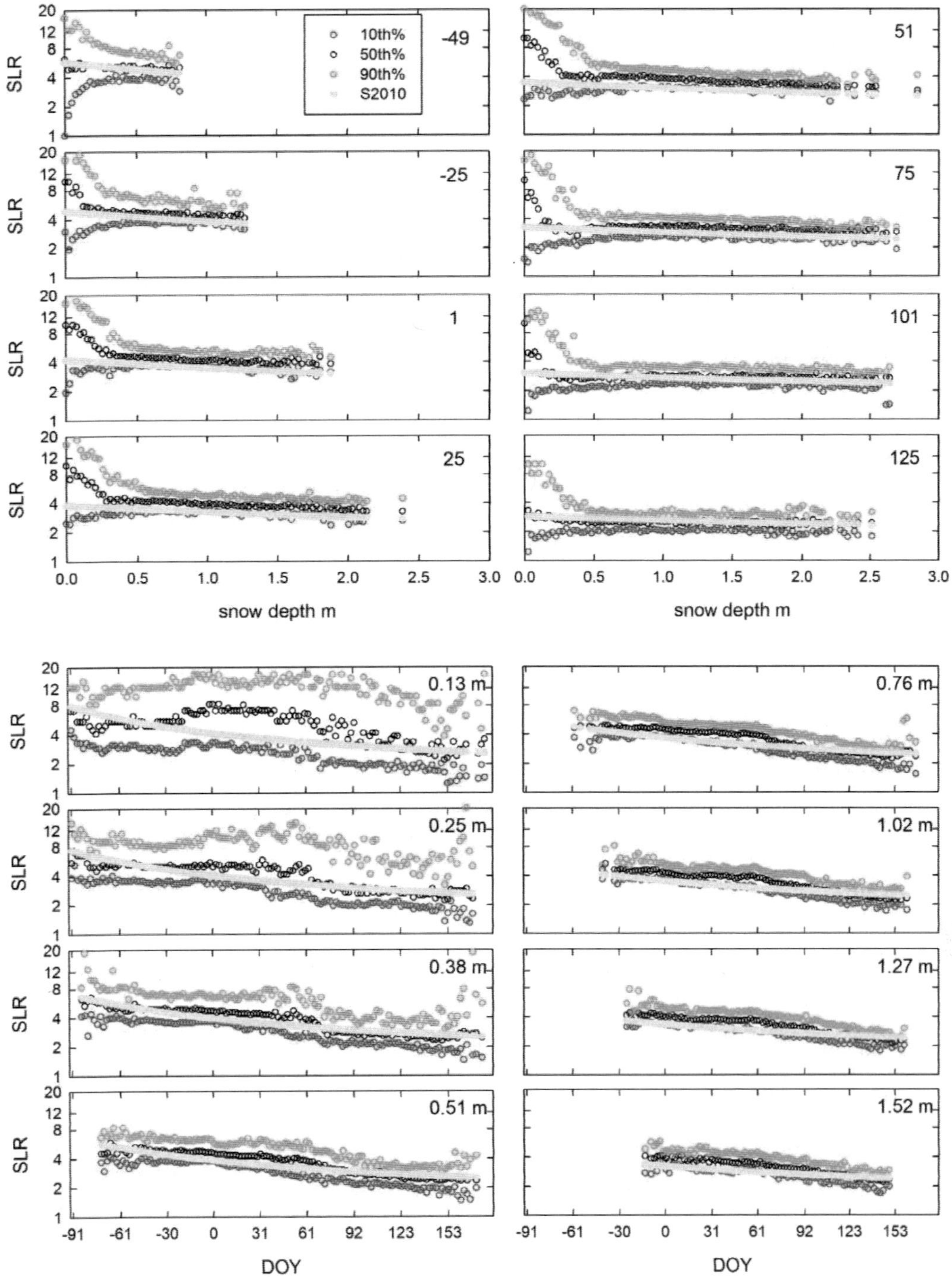

Figure 11e. SLR quantiles for GHCN data in the Prairie snow class: (top) as a function of snow depth for eight *DOY*, (bottom) as a function of *DOY* for eight snow depths. SLRs calculated from the S2010 fit are also shown.

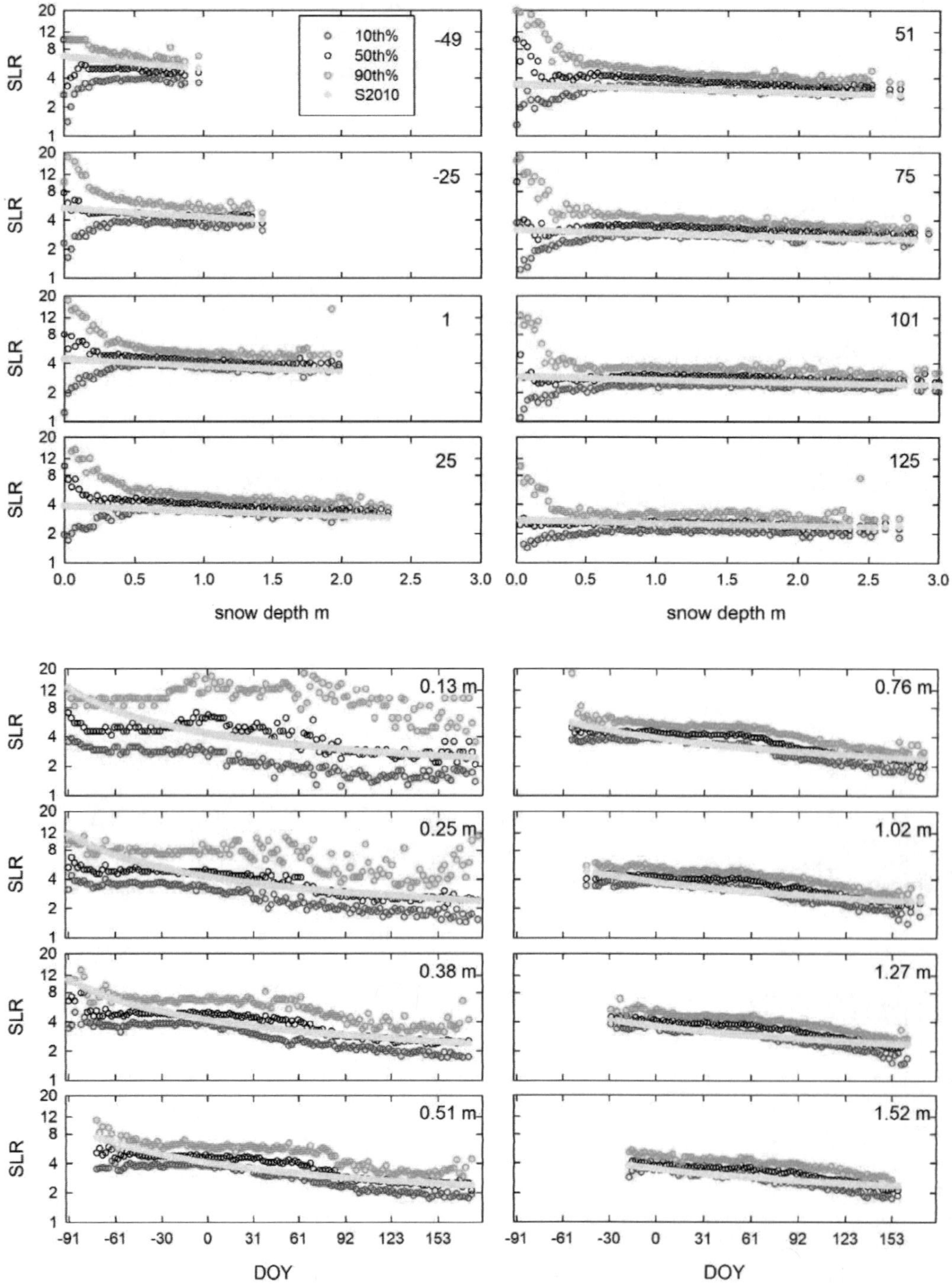

Figure 11f. SLR quantiles for GHCN data in the Alpine snow class: (top) as a function of snow depth for eight *DOY*, (bottom) as a function of *DOY* for eight snow depths. SLRs calculated from the S2010 fit are also shown.

class. Taking the inverse of the SLR quantiles in Table 8 to get snow density, the typical Taiga density is 0.17 g cm^{-3} with 80% of the values between 0.1 and 0.24 g cm^{-3}. Ephemeral snow densities are higher, with 80% of the values between 0.18 and 0.43 g cm^{-3} around a typical density of 0.29 g cm^{-3}.

Table 7. Median of the ratios of the 90th% and 10th% to the 50th% SLR for deep snow at the GHCN stations in three snow classes. The last row is the median of the ratio of the 50th% SLR to the S2010 fit.

Snow class	Maritime	Prairie	Alpine
Minimum depth	0.75 m	0.5 m	0.5 m
Ratio 90th% to 50th%	1.22	1.23	1.21
Ratio 10th% to 50th%	0.81	0.81	0.83
Ratio 50th% to S2010	1.03	1.16	1.13

Table 8. SLR quantiles for snow deeper than 0.15 m (6 inches) at the GHCN stations in the Tundra, Taiga, and Ephemeral snow classes.

Snow class	Tundra	Taiga	Ephemeral
Minimum depth	0.15 m	0.15 m	0.15 m
10th% SLR	4.2	4.2	2.3
50th% SLR	9.9	5.8	3.5
90th% SLR	14.9	10.	5.7

With the exception of the Tundra class, with data primarily from WBAN stations with too many SLR = 10, Figure 11 is dominated by reports from SNOTEL stations. The Maritime snow class, however, has both many cases and a significant fraction of data from other networks. To compare the SNOTEL SLRs with those computed from manual snow depth and SWE measurements, we plotted the SLR quantiles as a function of snow depth for this snow class separately for the SNOTEL and non-SNOTEL networks in Figure 12. Because all the CoCoRaHS, COOP, and WBAN station in this snow class are at low elevations, this comparison is limited to data from stations with elevations below 1000 m. This includes 99% of the 4746 non-SNOTEL Maritime stations and 17% of the 168 SNOTEL Maritime stations. There is somewhat deeper snow at the SNOTEL stations, and the 10th% and 90th% quantiles are closer together. The most obvious difference, however, is the good agreement between the GHCN 50th% SLRs and S2010 for the SNOTEL stations, while S2010 is close to the 10th% SLR at the non-SNOTEL stations. While this comparison between SNOTEL and non-SNOTEL SLRs is limited to stations at elevations below 1000 m in the Maritime snow class it is not clear whether the cause of the differences is the way snow depth and SWE measurements are made or the properties of the snow. These 29 SNOTEL stations, at an average elevation of 607 m, experience deeper snow than the 4696 non-SNOTEL stations, at an average elevation of 217 m. Perhaps the larger SLRs at the non-SNOTEL stations are due to the thinner snow pack. We looked at further refining the stations for this comparison to better match elevations, choosing the nine SNOTEL stations at the lowest elevations, or 309 non-SNOTEL stations at the highest elevations. This would reduce the number of samples in each snow depth-DOY bin so much that the quantiles would be meaningless.

SUMMARY AND CONCLUSIONS

The GHCN database provides valuable data on snow depth and snow water equivalent. The vast majority of these data pairs are from North America. The manual measurement protocols for snow depth and SWE for the CoCoRaHS, COOP, and WBAN networks are similar. The parameters are measured automatically at the remote SNOTEL stations. All networks report these data in inches, which are converted to millimeters in the GHCN database. We used simultaneous snow depth and SWE data from the stations in these networks to calculate the snow to liquid ratio. By computing the cumulative distribution of SLRs from each network, we determined that the reported SWE data are apparently sometimes calculated from the snow depth by dividing by 10 rather than measured. This has occurred more frequently at stations in the WBAN and COOP networks than at CoCoRaHS stations. Exact 10 SLRs occur about as frequently as expected at SNOTEL stations. Physically impossible SLR values, indicating snow densities greater than the density of bubble-free ice also occur, most frequently at stations in the COOP network. The GHCN QA algorithms check snow depth and SWE separately but do not test the resulting SLR. Users relying on the snow depth or SWE data in GHCN should consider calculating their ratio, when both parameters are available, as a further QA check.

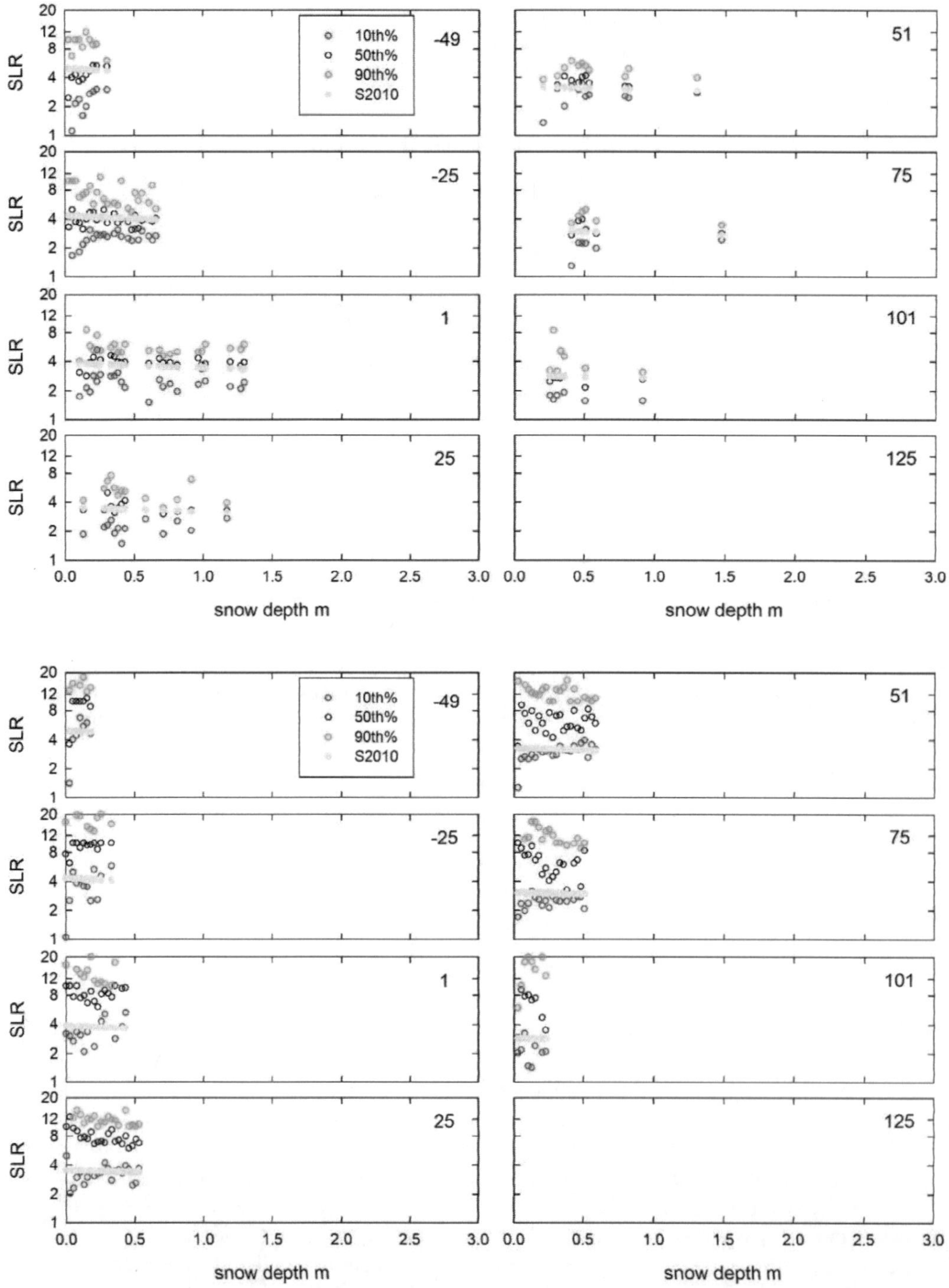

Figure 12. SLR quantiles, for GHCN data in the Maritime snow class at stations with elevation ≤ 1000 m, as a function of snow depth for eight snow classes; (top) SNOTEL data, (bottom) CoCoRaHS, COOP, and WBAN data. SLRs calculated from the S2010 fit are also shown.

We eliminated the snow depth-SWE pairs that failed the GHCN QA checks and assigned the data to the Sturm *et al.* (1995) snow classes. For each DOY-snow depth bin, we quantified the GHCN SLRs by the 10th%, 50th% (median), and 90th% quantiles to identify the center and range of the SLR distributions in each snow class. We compared these GHCN SLR quantiles to those calculated from the S2010 snow density fit.

Our results indicate substantial variation in SLR for snow depths less than 0.75 m, or less than 0.5 m for Alpine and Prairie snow. In the Tundra, Taiga, and Ephemeral snow classes, the snowpack tends to be shallow. There are excessive SLR = 10 in the Tundra snow class, associated with the large fraction of data from WBAN stations, so our results probably do not reflect actual snow properties at those stations. The SWE at GHCN stations in the Taiga and Ephemeral snow classes appears to be measured rather than calculated for most reports. Most (80%) of the Taiga SLRs are between 4.2 and 10, while the Ephemeral SLRs between 2.3 and 5.7 indicate denser snow with less density variation.

In deep snow in the Maritime, Prairie and Alpine classes, 80% of the snow measurements fall within about 20% of the median SLR, which varies with snow depth and *DOY*. The median is only slightly higher than the S2010 fitted SLR in the Maritime class but about 15% higher than the Prairie and Alpine S2010 SLR. The slopes with respect to *DOY* of the GHCN SLR quantiles increase at about *DOY* 61, likely caused by much less new snow and a densifying snow pack.

We used data from the Maritime snow class to try to compare manual and SNOTEL snow measurement. The data from stations at elevations below 1000 m show a significantly higher SLR at the non-SNOTEL stations. It is not clear if this difference is related to a systematic difference between manual and automatic snow depth and SWE measurements or to the lower elevations of the non-SNOTEL stations.

REFERENCES

Alcott TI, Steenburgh WJ. 2010. Snow-to-liquid ratio variability and prediction at a high-elevation site in Utah's Wasatch Mountains. *Weather and Forecasting,* **25**: 323-337.

Baxter MA, Graves CE, Moore JT. 2005. A climatology of snow-to-liquid ratio for the contiguous United States. *Weather and Forecasting,* **25**: 729-744.

Huntington TG. 2005. The density of falling snow in New England, 1949-2001. *Proceedings of 62nd Eastern Snow Conference*, Waterloo, Canada.

Jones KF, Daly SF. 2016. Effect of Arctic Amplification on design snow loads in Alaska, ERDC/CRREL MP-16-1, 56 pp.

Menne MJ, Durre I, Vose RS, Gleason BE, Houston TG. 2012a. An overview of the Global Historical Climatology Network-Daily Database. *Journal of Atmospheric and Oceanic Technology,* **29**: 897-910, doi:10.1175/JTECH-D-11-00103.1

Menne MJ, Durre I, Korzeniewski B, McNeal S, Thomas K, Yin X, Anthony S, Ray R, Vose RS, Gleason BE, Houston TG. 2012b. Global Historical Climatology Network - Daily (GHCN-Daily), Version 3.24 NOAA National Climatic Data Center. http://doi.org/10.7289/V5D21VHZ [9.4.2018]

Meyer JDD, Jin J, Wang S-Y. 2012. Systematic patterns of the inconsistency between snow water equivalent and accumulated precipitation as reported by the Snowpack Telemetry Network. *Journal of Hydrometeorology,* **13**: 1970-1976.

NOAA. 2017. Federal Meteorological Handbook No.1 Surface weather observations and reports, FCM-H1-2017.

NWS. 2014. Snow Measurement Guidelines for National Weather Service Surface Observing Programs, https://www.weather.gov/media/coop/Snow_Measurement_Guidelines-2014.pdf.

NWS. 2017. National Weather Service Manual 10-1315 (17 Oct 2017), Operations and Services, Surface Observing Program (Land), NDSPD 10-13, Cooperative Station Observations and Maintenance. https://www.nws.noaa.gov/directives/sym/pd01013015curr.pdf

Sturm M, Holgren J, Liston GE. 1995. A seasonal snow cover classification system for local to global applications. *Journal of Climate,* **8**: 1261-1283.

Sturm M, Taras B, Liston GE, Derksen C, Jonas T, Lea J. 2010. Estimating snow water equivalent using snow depth data and climate classes. *Journal of Hydrometeorology,* **11**: 1380-1394.

Documenting Winter Snow Accumulation and Ablation of a Shrub-Tundra Catchment using Unmanned Aerial Systems and *in situ* Observations

BRANDEN WALKER[1], BARUN MAJUMDER[1], EVAN WILCOX[1], BRAMPTON DAKIN[1], THOMAS MISZTELA[1], AND PHILIP MARSH[1]

ABSTRACT

Arctic tundra environments are characterized by a spatially heterogenous end-of-winter snow distribution resulting from wind transport and deposition. Large spatial variations in snow depth, density and snowpack microstructure result in localized concentrations of water storage across the landscape influenced by topography and vegetation cover. Understanding the distribution of snow across tundra environments is important as the snow accumulation typically accounts for over half of the annual precipitation and is the dominant driver of the hydrological system. Currently, our ability to accurately measure snow has proven difficult and traditional methods often fail to accurately represent small-scale variations in snow cover at catchment scales. Furthermore, the accumulation patterns at landscape scales are poorly documented resulting from technical and environmental limitations. In this study we document spatial variations in snow depth accumulation and ablation across a shrub-tundra catchment as part of the TVCSnow campaign from Trail Valley Creek, NWT. We applied Structure-from-Motion photogrammetry using a fixed-wing Unmanned Aerial System (UAS) resulting in high-resolution snow depth mapping (1 meter) at five key periods of snow accumulation and throughout the snowmelt period. In combination with aerial surveys, snow depth and water equivalent measurements were recorded across the winter accumulation period resulting in a detailed documentation of snow accumulation and ablation for various key landcover types. The ability to capture high-resolution spatio-temporal changes to tundra snow cover furthers our understanding of the relative importance of various land cover types on winter snow accumulation and ablation which has strong implications on the hydrological system during the spring freshet.

[1] Cold Regions Research Centre, Wilfred Laurier University, Waterloo, ON, Canada

Session #5: Snow Research to Operations

Session Chair: Dorothy Hall

NASA SnowEX 2019/20

CARRIE VUYOVICH[1], HP MARSHALL[2], CHRISTOPHER HIEMSTRA[3], LUDOVIC BRUCKER[4,5], KELLY ELDER[6], AND JERRY NEWLIN[7]

ABSTRACT

The NASA SnowEx Mission is a multi-year effort to evaluate and improve our ability to measure and monitor snow water equivalent (SWE) and other snow characteristics. The NASA SnowEx19/20 Campaign consists of coordinated airborne and field-based experiments in the Western U.S., from the fall of 2019, through the spring of 2020. This effort includes two major components: 1) a detailed experiment on Grand Mesa, Colorado, and 2) a time series experiment over 13 sites, spanning 5 states, with biweekly field and airborne observations. These observations are aligned with the SnowEx Science Plan (Durand *et al.*, 2018) and address gaps in snow estimation capabilities in various land cover types and snow classes (forest, mountain, prairie and maritime), and throughout the snow season (accumulation and melt), as well as characterize the snow surface energetics. The specific goals of these measurements are to: quantify accuracy and limitations of L-band InSAR retrievals of change in SWE, in preparation for NISAR; test and validate SWE retrieval from a multi-frequency radar and radiometer sensor package; test Ka-band InSAR for snow depth retrieval and quantify bias due to penetration; quantify the subpixel variability in thermal IR signatures, and the effect on coarse resolution spaceborne IR (GOES16); and integrate *in situ* and airborne data with modeling (e.g. THP16 SEUP, NOHRSC SNODAS). In this presentation we provide an overview of the SnowEx 2019/20 science objectives, experimental plan, and schedule.

---

[1] Hydrological Sciences Laboratory, NASA Goddard Space Flight Center, Greenbelt, MD, USA
[2] Department of Geosciences, Cryosphere Geophysics and Remote Sensing (CryoGARS), Boise State University, Boise, ID, USA
[3] U.S. Army Corps of Engineers, Engineering Research and Development Center, Cold Regions Research and Engineering Laboratory, Fairbanks, AK, USA
[4] Cryospheric Sciences Laboratory, NASA Goddard Space Flight Center, Greenbelt, MD, USA
[5] Universities Space Research Association, GESTAR, Columbia, MD, USA
[6] Fraser Experimental Forest, U.S. Forest Service, Ft. Collins, CO, USA
[7] Applied Technology Associates (ATA) Aerospace, Greenbelt, MD, USA

USDA Natural Resources Conservation Service Snow Survey and Water Supply Forecasting Program

MICHAEL L. STROBEL[1]

ABSTRACT

The Snow Survey and Water Supply Forecasting Program collects high elevation snow data in 13 states in the western U.S. and provides snowpack information, climatic data, and water supply forecasts.

The demographic, physical, and political landscape of the western U.S. is changing rapidly, and there is competition over water for irrigation, municipal and industrial customers, and in-stream uses, such as river-based recreation, fish and wildlife habitat, and hydroelectric power generation.

Extremes in the snowpack could result in less reservoir storage in warm, dry years, complicate reservoir regulation in cold, wet years, and cause extensive local and regional flooding. Earlier snowmelt, caused by warming conditions, increases the length of time between peak flows and summer water user needs, while a delayed snowmelt, caused by cool weather, shortens the melting season and produces potentially disastrous flooding. Drought throughout much of the western US and declining winter snowpacks have stressed hydrologic conditions and increased the risk of wildfires.

Water supply forecasts are used by: (a) irrigators for agricultural production needs; (b) Federal government in administering international water treaties; (c) State governments in managing intrastate streams and interstate water compacts; (d) municipalities in managing water supplies and drought; (e) reservoir operators; (f) Federal and State governments to mitigate flood damages; and (g) Federal and State governments to support fish and wildlife management activities. SNOTEL and SCAN networks provide information on soil moisture and soil temperature used by a wide range of agencies and organizations.

[1] USDA-NRCS National Water and Climate Center, Portland, OR, USA

Eastern-SNOW: A Coordinated Eastern United States Snow Observation Network

ELIZABETH A. BURAKOWSKI[1], ALIX CONTOSTA[1], MICHAEL DURAND[2], AND JENNIFER JACOBS[1]

ABSTRACT

In the eastern United States, snow is essential to ecosystem function, to economic prosperity of rural communities, and to water supplies of both urban and rural areas. The snowpack can also pose hazards when it causes flooding and infrastructure damage. Snow performs similar roles the western US, but while a robust network of 800+ automated and standardized stations provide a detailed climatology of western US snow, the eastern US lacks such a spatiotemporally rich network. The hundreds of COOP observer stations that provide the majority of snow depth data in the East suffer from inconsistencies in reporting standards, missing data, and poor measurement quality, limiting the ability of eastern US snow data to meet a variety scientific, policy, and management needs. Furthermore, few COOP observers report snow water equivalent (SWE), and only five automated Soil Climate Analysis Network (SCAN) sites in the eastern US provide automated SWE observations. Because most models simulate snow depth as SWE, the dearth of eastern US automated and standardized SWE records leaves researchers lacking requisite data for model validation that underlies weather forecasting, flood risk assessment, and future climate projections. Here, we report on the current inventory of eastern US snow records. We will use the inventory to identify data gaps, prioritize community research needs, and inform design of a robust, coordinated Eastern United States Snow Observation netWork (Eastern-SNOW) whose objectives are to: (1) establish a standardized database for all eastern US snow observations, (2) identify priority locations for new, automated snow sensor sites, and (3) streamline data collection and processing from these sites. Although the proposed Eastern-SNOW network will complement the existing SNOTEL network, it will serve the unique research, management, and policy needs of the eastern US, particularly within the context of a changing winter climate.

[1] Institute for the Study of Earth, Oceans, and Space, University of New Hampshire, Durham, NH, USA
[2] School of Earth Science, The Ohio State University, Columbus, OH, USA

Enhanced Monitoring of Snow Cover Extent Across Northern Hemisphere Lands

DAVID A. ROBINSON[1] AND THOMAS W. ESTILOW[1]

ABSTRACT

A satellite climate data record of Northern Hemisphere snow cover extent (SCE) has been generated using 20 years of Interactive Multisensor Snow and Ice Mapping System maps (September 1998- present). This daily 24-km resolution dataset is produced at the US National Ice Center. IMS mapping succeeded a 190-km resolution weekly product that NOAA generated from 1966-1999. Throughout the 50+ years of weekly and IMS mapping, trained analysts have primarily employed visible satellite imagery and interactive means of SCE mapping.

The new 24-km climatology provides a more detailed local and regional assessment of intra-seasonal and inter-annual SCE variability. Thus far, the period of record is too short to gain a strong perspective on change at these spatial scales. However, it can be used to gain an understanding of locations driving long-term changes that have been identified at the coarser scale for much longer (a downscaled weekly product continues to be generated based on IMS Monday maps). Presently, 1981-1999 weekly maps are being re-digitized at 24-km resolution. When this project concludes in 2020, there will be a 40-year weekly 24-km product available for longer term, albeit not daily, assessments.

These SCE products serve user communities interested in assessing climate variability and change, understanding the role of snow in the climate system, verifying snow as depicted in climate models, and for applied studies in water resources, energy, engineering and other fields.

[1] Rutgers University, Piscataway, NJ, USA

Challenges and Innovations to Operational Hydrologic Forecasting in Alaska

JESSICA CHERRY[1,2]

ABSTRACT

This presentation will review the National Weather Service's water mission, past, present, and future and describe specific cold season challenges in Alaska. Approaches to estimation and assimilation of snow conditions into current and future operational models will be described, along with other geoscience innovations of interest to the cryospheric science community. These include new approaches to both *in situ* monitoring and remote sensing in Alaska. Emerging needs driven by changing environmental conditions will be discussed. Ongoing collaborations with research partners will also be described, as well as research gaps and challenges where the APRFC is interested in new development. Finally, specific barriers to overcome in moving from research-to-operations will be discussed.

[1] Alaska Pacific River Forecast Center, National Oceanic and Atmospheric Administration, National Weather Service, Anchorage, AK, USA
[2] University of Alaska Fairbanks, Fairbanks, AK, USA

TO ERR IS HUMAN, TO FORGET IT WOULD BE DIVINE.

SNO-FOO 2019 – Alexandre Roy

The 2019 Sno-Foo award is presented to Dr. Alexandre Roy, Professor in the Département des sciences de l'environnement at the Université du Québec à Trois-Rivières. Alex wins the award while serving as the Past President of the Eastern Snow Conference.

After being "blind-sided" by Alex Roy in the 2018 Sno-Foo nomination regarding a Ford F150 door incident during winter fieldwork in New Hampshire, the record needed to be set straight at ESC 2019. In fact, some details omitted from the 2018 nomination, were sufficient for a proper nomination for Alex Roy. Those details are as follows: when an esteemed colleague called from the cabin door to ask if either Alex had the cabin keys, Alex Langlois opened the vehicle door to answer the question. Meanwhile and presumably aware that Alex L. had opened the vehicle door and was in the process of answering the question, Alex R. put the vehicle into reverse, the results of which are documented in the 2018 ESC Proceedings. Why did he reverse? Perhaps we will never know. Alex L. believed that opening the vehicle door should have been an obvious sign of "do not back up, wait". Apparently, Alex R. needed more detailed instructions. Apparently, the truck door was not as flexible as the retelling of history.

To make matters worse, it was discovered in 2019 that the engraving of the Sno-Foo Award trophy had not been completed recently. Traditionally, it is the responsibility of each year's winner to engrave their own name on it, but it is also the responsibility of the ESC President to remind the winner to do so. For failing to properly maintain the coveted Sno-Foo Award trophy and his role in the New Hampshire truck-gate, Alexandre Roy is a fitting winner.

Presented by Alexandre Langlois

Past Recipients

The Eastern Snow Conference annually bestows upon a distinguished snow scientist who, in striving for excellence in snow research, contributes to an event of notable humor the highly coveted Sno-Foo Award.

Year	Winner	Affiliation
2019	Alexandre Roy	Université du Québec à Trois-Rivières
2018	Alexandre Langlois	Université de Sherbrooke
2017	Michael Durand	The Ohio State University
2016	Roger De Roo	University of Michigan
2015	Kevin Côté	Université de Sherbrooke
2014	Dorothy Hall	NASA GSFC
2013	Benoit Montpetit	Université de Sherbrooke
2012	Don Pierson	NYC DEP
2011	Ken Rancourt	Mt. Washington Observatory
2010	Kyung-Kuk (Kevin) Kang	University of Waterloo
2009	Rob Hellström	Bridgewater State University
2008	Steven Fassnacht	Colorado State University
2007	The Group of 9[1]	
2006	Andrew Klein	Texas A&M University
2005	Claude Duguay	University of Alaska-Fairbanks
2004	Chris Derksen	Meteorological Service of Canada
2003	Miles Ecclestone	Trent University
2002	Danny Marks	U.S.D.A.
2001	Brenda Toth	University of Saskatchewan
2000	Mauri Pelto	Nichols College
1999	Ross Brown	Meteorological Service of Canada
1998	Mary Albert	CRREL
1997	Jean Stein	Université de Laval
1996	Colin Taylor	Trent University
1995	Mike Demuth	NHRI
1994	Bert Davis	CRREL
1993	John Pomeroy	NHRI
1992	Tom Niziol	National Weather Service
1991	Terry Prowse	NHRI
1990	Kersi Davar	University of New Brunswick
1989	Gerry Jones	INRS-EAU
1988	Robert Sykes	SUNY
1987	John Metcalfe	Meteorological Service of Canada
1986	Peter Adams	Trent University
1985	Don Wiesnet	National Weather Service
1984	Barry Goodison	Meteorological Service of Canada

[1] Jimmy MacDonald (University of Saskatchewan), Bill Floyd (University of British Columbia), Chris DeBeer (University of Saskatchewan), Wendell Koning (Government of Alberta), Jaime Hood (University of Calgary), Danika Muir (University of Calgary), John Jackson (University of Calgary), Sarah Forte (University of Calgary), Masaki Hayashi (University of Calgary)

LIST OF ATTENDEES

Rosa Affleck	*Cold Regions Research and Engineering Laboratory*
Jawairia Ahmad	*University of Maryland*
Stefanie Arndt	*Alfred Wegener Institute*
Ian Baker	*Dartmouth College*
Samuel Baldwin	*Cold Regions Research and Engineering Laboratory*
Maxime Beaudoin-Galaise	*Université Laval*
James Brylawski	*GEONOR, Inc.*
Elizabeth Burakowski	*University of New Hampshire*
Jessica Cherry	*National Weather Service*
Krystopher Chutko	*University of Saskatchewan*
Guillaume Couture	*Université de Sherbrooke*
Derrill Cowing	*USGS (Retired)*
Roger De Roo	*University of Michigan*
Elias Deeb	*Cold Regions Research and Engineering Laboratory*
Michael Durand	*The Ohio State University*
Jonathan Edwards-Opperman	*U.S. National Ice Center*
Bruce Elder	*Cold Regions Research and Engineering Laboratory*
Lauren Farnsworth	*Cold Regions Research and Engineering Laboratory*
Bart Forman	*University of Maryland*
James Foster	*NASA Goddard Space Flight Center (Emeritus)*
Barry Goodison	*Environment and Climate Change Canada (Retired)*
Carl Green	*Cold Regions Research and Engineering Laboratory*
Grant Gunn	*Michigan State University*
Dorothy Hall	*University of Maryland*
Steven Hall	*Gradient Wind Engineering, Inc.*
Adam Hunsaker	*University of New Hampshire*
Jennifer Jacobs	*University of New Hampshire*
Kathy Jones	*Cold Regions Research and Engineering Laboratory*
DK Kang	*NASA Goddard Space Flight Center*
Eric Kelsey	*Mount Washington Observatory / Plymouth State University*
Edward Kim	*NASA Goddard Space Flight Center*
Joshua King	*Environment and Climate Change Canada*
John Kochendorfer	*National Oceanic and Atmospheric Administration*
Daniel Kramer	*Université de Sherbrooke*
Alex Langlois	*Université de Sherbrooke*
Theodore Letcher	*Cold Regions Research and Engineering Laboratory*
James Lever	*Cold Regions Research and Engineering Laboratory*
Adam LeWinter	*Cold Regions Research and Engineering Laboratory*
Yuan Li	*Dartmouth College*
Zhibang Lv	*University of Saskatchewan*
Victoria Ly	*University of Washington*
Iona MacDonald	*Tinicum Elementary School*

Mariah Matias	*Lehigh University*
Alex Mavrovic	*Université du Québec à Trois-Rivières*
Cooper McCann	*NWB Sensors, Inc.*
Rachel McCrary	*National Center for Atmospheric Research*
Terry Melendy	*Cold Regions Research and Engineering Laboratory*
Julien Meloche	*Université de Sherbrooke*
Michelle Michaels	*Cold Regions Research and Engineering Laboratory*
Benoit Montpetit	*Environment and Climate Change Canada*
Blaine Morriss	*Cold Regions Research and Engineering Laboratory*
Mohammad Mousavi	*University of Michigan*
Carrie Olheiser	*National Weather Service / National Center for Atmospheric Research*
Jongmin Park	*University of Maryland*
Julie Parno	*Cold Regions Research and Engineering Laboratory*
Mauri Pelto	*Nichols College*
Amandine Pierre	*Université Laval*
Chris Polashenski	*Cold Regions Research and Engineering Laboratory*
Joan Ramage	*Lehigh University*
Kenneth Rancourt	*Mount Washington Observatory*
George Riggs	*Science Systems and Applications, Inc.*
David Robinson	*Rutgers University*
Alexandre Roy	*Université du Québec à Trois-Rivières*
Alain Royer	*Université de Sherbrooke*
Sally Shoop	*Cold Regions Research and Engineering Laboratory*
Paul Siqueira	*University of Massachusetts*
Craig Smith	*Environment and Climate Change Canada*
Cob Staines	*University of Saskatchewan*
Mike Strobel	*National Water & Climate Center*
Alison Thurston	*Cold Regions Research and Engineering Laboratory*
Marissa Torres	*Cold Regions Research and Engineering Laboratory*
Leung Tsang	*University of Michigan*
Samuel Tuttle	*Mount Holyoke College*
Céline Vargel	*Université de Sherbrooke*
Carrie Vuyovich	*NASA Goddard Space Flight Center*
Branden Walker	*Wilfred Laurier University*
John Woods	*U.S. National Ice Center*
Gaohong Yin	*University of Maryland*
Jiyue Zhu	*University of Michigan*